HOMO

the mo
is a mi
Isaac
the human being from
comparing and contrasting his development
with that of other creatures.
Then chapter by chapter,
with a lively style and easy informality,
he details the intricate structure
and function of the human body,
male and female.

Mr. Asimov describes the skeletal framework,
the muscles, the system of blood vessels,
the digestive apparatus, and
the heart, liver, lungs, kidneys—
organs that fire and fuel the living body.
Last of all the author describes
the sheath of sensitive skin,
which covers the whole structure,
and the genitalia, the organs of reproduction,
which assure the regeneration of
the species.

Special features of
this volume
are aids to pronunciation,
derivations of specialized terms,
and clarifying illustrations by
the well-known medical illustrator
Anthony Ravielli.

MENTOR Books of Special Interest

(0451)

☐ **THE HUMAN BRAIN: Its Capacities and Functions by Isaac Asimov.** A remarkable, clear investigation of how the human brain organizes and controls the total functioning of the individual. Illustrated by Anthony Ravielli. (619013—$2.25)

☐ **THE HUMAN BODY: Its Structure and Operation by Isaac Asimov.** A superbly up-to-date and informative study which also includes aids to pronunciation, and derivations of specialized terms. Drawings by Anthony Ravielli. (621166—$2.95)

☐ **THE CHEMICALS OF LIFE by Isaac Asimov.** An investigation of the role of hormones, enzymes, protein and vitamins in the life cycle of the human body. (620372—$1.95)

☐ **THE DOUBLE HELIX by James D. Watson.** A "behind-the-scenes" account of the work that led to the discovery of DNA. "It is a thrilling book from beginning to end—delightful, often funny, vividly observant, full of suspense and mounting tension . . . so directly candid about the brilliant and abrasive personalities and institutions involved . . ."— *New York Times.* Illustrated. (622804—$2.95)*

☐ **INSIDE THE BRAIN by William H. Calvin, Ph.D. and George A. Ojemann, M.D.** In the guise of an operation, we are given a superbly clear, beautifully illustrated guided tour of all that medical science currently knows about the brain—including fascinating new findings. (620526—$2.95)

*Price is $3.50 in Canada

Buy them at your local bookstore or use this convenient coupon for ordering.

THE NEW AMERICAN LIBRARY, INC.,
P.O. Box 999, Bergenfield, New Jersey 07621

Please send me the books I have checked above. I am enclosing $_____ (please add $1.00 to this order to cover postage and handling). Send check or money order—no cash or C.O.D.'s. Prices and numbers are subject to change without notice.

Name_____

Address_____

City _____ State _____ Zip Code _____
Allow 4-6 weeks for delivery.
This offer is subject to withdrawal without notice.

THE HUMAN BODY
Its Structure and Operation

by ISAAC ASIMOV

Illustrated by Anthony Ravielli

A MENTOR BOOK

NEW AMERICAN LIBRARY

NEW YORK AND SCARBOROUGH, ONTARIO

COPYRIGHT © 1963 BY ISAAC ASIMOV

All rights reserved, including the right to reproduce
this book or parts thereof in any form. For information
address Houghton Mifflin Company, 1 Beacon Street,
Boston, Massachusetts 02107.

Model of human body reproduced
with permission of Merck & Co., Inc.

This is an authorized reprint of a hardcover edition
published by Houghton Mifflin Company.

MENTOR TRADEMARK REG. U.S. PAT. OFF. AND FOREIGN COUNTRIES
REGISTERED TRADEMARK—MARCA REGISTRADA
HECHO EN CHICAGO, U.S.A.

SIGNET, SIGNET CLASSIC, MENTOR, PLUME, MERIDIAN AND NAL
BOOKS are published in the United States by
New American Library,
1633 Broadway, New York, New York 10019,
in Canada by The New American Library of Canada Limited,
81 Mack Avenue, Scarborough, Ontario M1L 1M8

13 **14** 15 16 17 18 19 20 21

PRINTED IN THE UNITED STATES OF AMERICA

CONTENTS

THE HUMAN BODY

1

OUR PLACE

In writing a book about the human body there is the great advantage that all the readers know what a human body is. They can recognize it at a glance and distinguish it, let us say, from a rock or from a cypress tree, from an oyster or frog, from a dog, even from a chimpanzee. What is more, we all know the prominent external features of the human body to begin with and even have some idea as to its internal features. We know something about its more obvious ways of functioning.

In a sense, this is a disadvantage too, for there is always the temptation to assume all this knowledge and to plunge into the description and discussion of the human body without as much as a sideward glance. And yet man does not exist in isolation: he is a small part of the domain of life and, in a still larger sense, is an even smaller part of the totality of all things.

There is profit to be gained, now and then, in viewing the structure and workings of the human body, not in isolation, but against this background of life and the universe, and it will be helpful therefore if we pretend an ignorance we do not really have. Let's concern ourselves with the definition of what, exactly, we mean by the human body.

One way of doing this logically and systematically is to try to divide all things into two or more classifications by making some reasonable distinction among them, and then to place man in one of those classifications. Concentrating on the classification into which we have placed him, we can divide that further by making finer distinctions, place him anew, and so proceed as far as is necessary for the purposes of this book.

To begin with, for instance, we can say that a rock is not a man. A rock does not feed, grow, and multiply; it does not sense its environment and respond to it in an adaptive way (that is, in a way designed to protect its existence). A man, however, does all these things. In this way, we distinguish not only a rock from a man, but all nonliving things from all living things. We make the first clear-cut distinction leading to the proper placing of man by stating that he is alive.*

If we restrict ourselves next to living things, we can go on to say with great confidence that it is easy to distinguish a cypress tree or a cactus from man. The former are rooted in the ground and are incapable of rapid voluntary motion. Important portions of their surface are green. A man, on the other hand, is not rooted, can move rapidly, possesses no green parts of any significance, and so on. A great many other distinctions can easily be made which will divide living things into the two grand divisions of the *Plant Kingdom* and the *Animal Kingdom,* with ourselves clearly a member of the latter. (Some biologists define a third and even a fourth kingdom, but these involve microscopic creatures only and need not concern us.)

When we confine ourselves to animal life and try to restrict man into a still narrower group, things cease to be quite so easy. It is almost automatic to try to make the kind of distinction that will divide a heterogeneous group

---

* The difference between life and nonlife, under the circumstances ordinarily encountered, is sufficiently great for even the casual observer to find words sufficiently exact to make the distinction easily. As we study simpler and simpler forms of life, however, matters become less clear and, at the simplest levels, the distinction becomes difficult to express. This book is not the place to work out the necessary concepts and phraseology, but if you are curious, you will find this in another book of mine, *Life and Energy* (1962).

in two. That is the simplest form of classification and I
have already done this twice: nonlife and life; plant and
animal. There were bound to be attempts to continue this
binary classification.

The Greek philosopher Aristotle (who lived in the 4th
century B.C.) classified all animals into those with blood
and those without blood, and man, of course, came into
the former group. Yet almost all animals have blood of
one sort or another, and even if we restrict ourselves to
red blood (which is what Aristotle undoubtedly meant)
the division is erratic and gives rise to two quite heterogeneous groups, neither of which can be easily studied
as a whole.*

In modern times another attempt was made, one closer
to the mark. The French naturalist Jean Baptiste Lamarck,
in 1797, divided animals into *vertebrates* and *invertebrates*.
The vertebrates include those animals possessing a backbone made up of a series of individual bones called
"vertebrae." The invertebrates, very naturally, include all
other animals. According to this system man would be a
vertebrate.

In one respect this is good. The vertebrates make up a
comparatively close-knit group of animals. In another
respect it is not so good, for the invertebrates are such
a wide range of animals with such fundamental differences
among themselves that they cannot be studied as a unified
whole. To the average man, far more interested in himself
than in any other animal, the term invertebrate is good
enough, however, and Lamarck's division is still widely
used in popular writing. After all, among the invertebrates
are bugs, worms, jellyfish, starfish, and other creatures
of little moment to the layman and they can be dismissed
easily enough. To the zoologist, intent on classifying *all*
animal life in a reasonable fashion, matters cannot be
dismissed so easily. It early became evident that no simple

* I should pause here to point out that in drawing distinctions and
making classifications, mankind is usually imposing artificial divisions
of his own choosing upon a universe that is, in many ways, "all one
piece." The justification for doing so is that it helps us in our attempted
understanding of the universe. It breaks down a set of objects and
phenomena too complex to be grasped in its entirety into smaller realms
that can be dealt with one by one. There is nothing objectively "true"
about such classifications, however, and the only proper criterion of
their value is that of their usefulness.

line could be drawn clear across the Animal Kingdom. Rather a number of lines must be drawn. The first to do so with marked success was the 18th-century Swedish botanist Carolus Linnaeus. In 1735 he published a book in which the forms of life were classified into divisions and subdivisions along lines that have been broadly adhered to ever since.

THE PHYLA

It was not Linnaeus, however, who gave the modern name to the broadest groups that can be included within the Animal Kingdom. This feat fell to the French naturalist, Georges Léopold Cuvier, a contemporary of Lamarck. In 1798 he divided the Animal Kingdom into four major branches and called each of these branches a *phylum* (fy′lum; plural, *phyla*), from a Greek word meaning a tribe or race.*

As time went on and zoologists studied the Animal Kingdom more closely and in greater detail, Cuvier's four phyla seemed overmodest. At the present time, there are roughly twenty phyla accepted. I say "roughly," because all such classifications, being man-made, depend upon the individual judgment of the classifier. There are borderline cases where a group of animals is classed with a particular phylum by one investigator but is judged sufficiently different by another to be given a phylum all to itself.

Each phylum (at least in the intent of the classifier) includes all animals that follow a certain basic plan of structure that is quite different, in important ways, from the one followed by all other animals. The best way to explain what is meant by that is to give some examples. This will tell us, in the end, what the basic plan of structure of the human body (as well as the bodies of related

* I intend, whenever it seems advisable, to give the pronunciation and derivation of technical terms. It seems to me this ought to help invest an otherwise nutcrakerish term with familiarity and perhaps lessen its terrors. Thus, a phylum may mean nothing to us, but we all know what a tribe is, and once we can pronounce phylum there is no further reason to be concerned about it. Since most scientific terms are derived from either Latin or Greek, I shall save space by putting the literal meaning in parenthesis with an L or G added: *phylum* ("tribe" G). Where other languages are involved, or where the derivation is of particular interest, I shall of course go into greater detail.

animals) is. Almost as important, it will tell us what the basic plan of structure of the human body is not. Finally, it will give us a framework to which we can refer later in the book from time to time.

The phylum *Protozoa* (proh-toh-zoh′ah; "first animals" G) includes all animal organisms consisting of a single cell. (I'll have more to say about cells later, but I shall assume now that you have heard enough about cells for the statement to have meaning for you.) That the phylum consists only of single-celled organisms is quite distinctive, for all the remaining animal phyla are made up of organisms that consist of a number of cells (*multicellular organisms*).

Consider, as another example, a pair of phyla— *Brachiopoda* (bray′kee-op′oh-duh; "arm-leg" G) and *Mollusca* (muh-lus′kuh; "soft" L).* The animals of both phyla possess hinged double-shells constructed of calcium carbonate (limestone). In this they are unique. To be sure, there are creatures outside these phyla that have calcium carbonate shells—as, for instance, the corals. The calcium-carbonate shells of corals and other outsiders, however, are in one piece and are not composed of two hinged halves. You might wonder why, then, the animals with hinged double-shells of similar chemical composition are nevertheless divided into two phyla. Well, among the mollusks one half of the shell is formed beneath the animal and the other half above, and the two are generally unequal in size. Among the brachiopods the halves of the shell are formed to the left and right of the animal and are roughly equal in size.

These are by no means the only important differences among the two groups, but even alone they would be considered sufficient by zoologists to make two phyla mandatory. (Just to indicate that matters are not really simple, there are mollusks with more than two shells, with only

---

* Each name is poor if the meaning of the word is taken literally. The name Brachiopoda was suggested because the naturalist who first studied the creatures assumed that certain structures served them as both arms and legs. This proved not to be so. The animals in the phylum Mollusca are not particularly soft. In fact, most have hard shells. Internally, of course, they are soft, but no softer than other animals. Nevertheless, as long as zoologists agree on a particular name it serves a useful function even when the literal meaning is wrong. And the derivation is always of historical interest, too.

one shell, and with no shell at all. These nevertheless belong in the phylum on the basis of other qualifications.)

The phylum *Echinodermata* (ee-ky'noh-dur'muh-tuh; "spiny-skin" G), although among the more complex of the phyla as far as most of its structure is concerned, can be differentiated from all other phyla of similar complexity by its possession of radial symmetry. This is a primitive characteristic; one, in other words, that is ordinarily associated with quite simple organisms.

Most phyla have bilateral symmetry. That is, an imaginary plane can be drawn through the body which will divide it into two halves that are mirror images of each other. There is thus a distinct left and right, and, if no other plane of symmetry can be drawn, there is also a distinct front and back, or, if you prefer, a distinct head and tail. The human being is clearly a member of a bilaterally symmetric phylum. His paired organs—eyes, ears, nostrils, arms, legs, and so on—are placed symmetrically on either side of a midplane running from head to feet. His single organs—nose, mouth, navel, anus, and so on—are placed at the midplane.

In radial symmetry no such unique plane can be drawn. Instead, there is a central point around which structures radiate. In the case of the echinoderms, there are usually five equivalent structures radiating out from the center, a fact that is most evident in the starfish, which is the best-known echinoderm.

Two phyla, *annelida* (a-nel'i-duh; "little ring" L) and *Arthropoda* (ahr-throp'oh-duh; "jointed feet" G), display a plan of structure that is shared by a third phylum, which I shall discuss later. This plan is that of segmentation; the organism is divided into a number of sections of similar

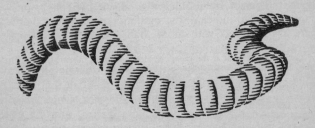

structure, something like a railroad train made up of similar passenger cars. This is obvious in the best known of the annelids, the earthworm, where the creature is divided into clearly marked off sections. The result resembles successive little rings of tissue and it is from this that the name of the phylum is derived. In some arthropods, such as centipedes, the existence of segments is as clear as it is in the earthworm. In others, the segmentation may be rather masked but it still shows up as a repetition of structures along the length of a creature; as, for example, the series of legs of the lobster. The two phyla, although sharing this very fundamental characteristic of segmentation, are clearly distinguished from each other by the fact that the annelids have no hard tissues, whereas the arthropods have a shell. (There are also additional distinguishing qualifications, of course.)

Nor is the arthropod shell to be confused with those of either the mollusks or the brachiopods. The arthropod shell is made of *chitin* (ky'tin), from a Greek word for a type of garment. Chitin is an organic substance built up of complex sugar molecules. It is tough, light, and flexible, whereas the stony calcium carbonate of the other phyla is hard, heavy, and brittle.

But the human being is also segmented. It is not as obvious in the human as in the earthworm or in the lobster, but he is. Does this make him a member of either Annelida or Arthropoda? Not necessarily. As we have seen in the case of Mollusca and Brachiopoda, resemblance in one respect is not enough. The human being, in addition to being segmented, has an elaborate internal skeleton. This no annelid and no arthropod possesses, and the difference is sufficiently fundamental to debar the human being from membership in either of these phyla.

THE DEVELOPMENT OF PHYLA

It is the opinion of biologists that the various phyla have not been independent throughout the past but that all have been descended from a common ancestor. Unfortunately, the order in which the phyla have arisen, and the exact manner in which one developed from another are not definitely known, though reasonable speculations may be and have been advanced.

The past history of living things is obtained most clearly from fossil records, the petrified remnants of creatures long dead, uncovered in deep-lying rocks. The earliest fossils that clearly show animal structure are found in the rocks of the Cambrian Period (named for Cambria, the Roman name for Wales, where these rocks were first studied). The rocks of the Cambrian Period date back half a billion years and more, and at that time all the phyla but one were already clearly established in advanced form so that the connections among them were no longer obvious.

Decisions as to the detailed evolution of the phyla must be based, therefore, on indirect information. For instance, since the arthropods and annelids both possess segmentation, and since the arthropods are, on the whole, the more complex in structure, it sounds reasonable to suppose that long ago, more than a half-billion years ago certainly, a group of annelids developed a chitinous shell and became the first arthropods.

This assumption, reasonable in itself, is strengthened by the existence, today, of an animal called *peripatus* (pee-rip'uh-tus; "walking about" G, because of the quickness of its scurrying). It is classified as an arthropod but is clearly the most primitive of the arthropods and possesses some characteristics that zoologists would ordinarily expect to find among the annelids. It is a "missing link," therefore; the descendant of a line of creatures that may have been annelids once and had not yet become complete arthropods.

What would interest zoologists most of all, of course, would be the establishment of a clear line of descent for the phylum that includes man. This phylum I have not (deliberately) yet mentioned.

The phyla I have mentioned clearly differ from man in very fundamental ways and he cannot be included in any of them. Unlike protozoans, we are made up of many cells. Unlike brachiopods, mollusks, and arthropods, we have no shell of any kind. Unlike annelids, we do have hard tissues. Unlike echinoderms, we are bilaterally symmetrical.

And yet our phylum must have been developed from

one of the others. Our lack of knowledge as to how this occurred is particularly frustrating since it happens, so to speak, in front of our eyes. I said that by the Cambrian Period all the phyla but one had already been established. That one, not yet established, is our own; and its manner of establishment has left no record that has yet been found. By the time very shortly after the Cambrian Period when the first fossils of animals of our own phylum make their appearance, they are already as far advanced as many creatures now living. The origin is lost or, at best, is as yet uncovered.

Nevertheless, all hope is not gone. There is indirect evidence. In the muscles of animals of our phylum, there is a compound called *creatine phosphate* (kree'uh-teen) which plays an important part in the chemistry of muscle contraction. In all other phyla (with one exception) creatine phosphate does not exist, but a similar role is taken by an allied compound called *arginine phosphate* (ahr'jih-neen). The one exception is the echinoderm phylum, some members of which also use creatine phosphate.

This is curious. Can we be descended from the echinoderms? Their radial symmetry makes them seem to differ from ourselves more than the animals of almost any other phylum.

Then, too, our phylum consists of segmented animals. To be sure, this is well masked in most cases, but you can verify this on your own body by running your hand down your backbone. You will clearly feel a series of similar bones, one to each segment, a repetition of structure as characteristic of segmentation as the repeated tissue rings of an earthworm or the repeated legs of a lobster. Since this is so, is it possible that our phylum, like that of the arthropods, is descended from the annelids?

But similarities cannot always be used to argue descent. It often happens in the course of evolutionary development that two quite different groups develop marked similarities even though they are not closely related. Thus, whales have developed a fishlike form but they are far more closely related (on the basis of other criteria) to man than to fish. Again, bats have developed wings but are far more closely related to man than to birds. This development of similarities on the part of animals not closely related (usually through the pressure of exposure to similar environments) is called *convergence*.

It may be then that man's phylum is descended from the annelids and that the use of creatine phosphate by both ourselves and some of the echinoderms is an example of convergence. On the other hand, we may be descended from the echinoderms and the existence of segmentation in ourselves as well as in annelids and arthropods may be the result of convergence. Or we could be descended in another fashion altogether, and creatine phosphate and segmentation may represent convergence all around.

Fortunately, there is further evidence to help us decide.

It often happens that the very early stages of the development of an individual animal show structures that resemble those of far distant ancestors. For instance, even the most advanced multicellular animals begin life as a single cell, which can be taken to signify the ultimate descent of all the multicellular phyla from the protozoa.

As this single cell divides into numerous daughter cells, the mass of cells eventually forms a cup-shaped structure consisting of two layers. The outside layer is the *ectoderm* (ek'toh-durm; "outer skin" G) and the layer facing the inside of the cup is the *endoderm* (en'doh-durm; "inner skin" G). There is a phylum of creatures with bodies that consist essentially of elaborations of this cup-shaped structure. This phylum is *Coelenterata* (see-len'tur-ay'tuh; "hollow gut" G, because digestion takes place in the interior of the cup, which thus becomes a gut).

In all phyla more complicated than the coelenterates, however, a third layer of cells is formed, lying between the two original layers and therefore called the *mesoderm* (mee'soh-durm; "middle skin" G). In some phyla the mesoderm arises at the point of junction of ectoderm and endoderm, and in others it arises at several different points of the endoderm only.

This difference in the manner of origin of the mesoderm is felt to be extremely fundamental by zoologists. It seems logical to suppose that from primitive two-layered coelenterates, a billion years ago or more, two new phyla arose, each of which independently developed the mesoderm in its own fashion. From each of these original three-layered phyla, a number of modern phyla developed.

The phyla containing mesoderm are therefore divided into two superphyla, each representing a line of common descent far far back in time.

As it happens, the echinoderms and the annelids, the two candidates for the honor of representing our ancestry, are in different superphyla which, in fact, are named for them. In the echinoderm superphylum, which is much the smaller of the two, the mesoderm develops from several points in the endoderm. In the annelid superphylum, the mesoderm develops from the ectoderm-endoderm junction.

It is a relatively easy task now to decide to which of the superphyla our own phylum belongs by studying the manner in which the mesoderm develops. The answer is, quite clearly and definitely, that our phylum belongs to the echinoderm superphylum. Of all the phyla, then, Echinodermata must be most closely related to our own.

THE CHORDATES

But what, then, of radial and bilateral symmetry?

A clue can be obtained from the young of the echinoderms. In the case of a number of animals, the creature, as it hatches from the egg, has a structure quite different from that of the adult. In the course of its life it undergoes a radical change in order to became an adult. The most familiar example is the caterpillar that becomes a butterfly.

Such a young form, differing radically from the adult, is a *larva,* which is a Latin word for a kind of ghost. Just as a ghost arises from a man but does not have the form and structure of a man, so a larva arises from the eggs laid by the mother but lacks the form and structure of the mother. Sometimes (but not always) the larva will possess structure and function that we have reason to believe were characteristic of the creatures from which it descended, while the adult form is a later specialization.

For instance, many of the echinoderms, brachiopods, and mollusks spend most of their life fixed to some one position or are at best capable of but slow movement. The larval form of such creatures is, however, capable of free motion; and this is useful—it can select a spot for itself upon which to settle down. Were it as motionless as its parent, all the offspring would grow up about the parent and the competition for food might insure the death of all. It is reasonable to suppose, then, that these sessile creatures were descended from free-moving ancestors and to search in the larvae for possible other characteristics of those ancestors.

Well, the echinoderm larvae are not only capable of free motion; they are also bilaterally symmetrical. It is only after conversion into the adult form that radial symmetry appears. The radial symmetry would thus appear to be a secondary development, which may not have existed at all in the very earliest days of the echinoderms.

In fact, we can imagine that when the echinoderm superphylum arose from the primitive coelenterates, two general types arose. One of them developed radial symmetry and gave rise to the modern echinoderms and the other developed certain distinctive features that were characteristic of no other phylum, thus developing into creatures that were not echinoderms at all. The distinctive features were three in number (excluding the retention of bilateral symmetry, which is not distinctive but is shared with many other phyla): These three distinctive features are worth attention because remnants are retained in all members of the phylum, including man. In other words, we will be talking about structures we possess at least in rudimentary form and during at least part of our lives.

First, the creatures of the new phylum possessed a hollow nerve cord that ran along the back of the organism —a *dorsal cord* ("back" L). In all other phyla the central nerve cord, if it existed at all, was solid and ran along the abdomen—a *ventral cord* ("belly" L).

Second, the creatures of the new phylum possessed an internal rod made of a tough, light, and flexible gelatinous substance. Such internal stiffening does not exist in other phyla except for a gristle-like substance in the most advanced mollusks. Even there it does not exist in the form of a rod. Because, in its most distinctive form, the gelatinous rod runs the length of the animal just under the dorsal nerve cord, the rod is called the *notochord* (noh'toh-kord; "back-string"G).

Third, the creatures of the new phylum possess a throat which is perforated by a number of *gill slits*. As water passes into the mouth and out these slits, food can be strained.

Any one of these three unique characteristics is sufficient to mark off a separate phylum, and it is to this phylum that we belong. The phylum is named after the presence of the notochord and is therefore *Chordata* (kawr-day'tuh).

The original chordates are lost (perhaps irrevocably) in the past, as are the original echinoderms. All we have today are living specimens of each phylum, specimens that have been developing through hundreds of years of time and have lost all apparent similarity. Yet there are examples of primitive chordates even today that have not altogether lost touch with the echinoderms. Biologists are particularly interested in them not only for their own sake but because they outline what may well have been the course of evolution from some primitive sea urchin to the complex group of animals of which man is a member. Thus, to begin with, there is a sea-dwelling wormlike creature, discovered about 1820, with a head that ends in a proboscis shaped vaguely like a tongue or an acorn. Behind this is a collar-like structure that seems to resemble a barnacle. It is called *balanoglossus* (bal'an-oh-glos'us; "barnacle tongue" G). The interesting thing about the balanoglossus, wormlike though it seems, is that behind the collar is a series of gill slits, a fact which alone spells "chordate." Furthermore, in the region of the collar there is a definite hollow dorsal nerve cord and, sticking into the proboscis, a short length of stiffening material that seems to be a piece of notochord.

This and a few allied species represent the most primitive chordates known. And the crucial point is that the larva of the balanoglossus is so like the larvae of the echinoderms that when the balanoglossus larva was first discovered it was classified as an echinoderm. Surely the proof of our echinoderm descent is overwhelming.

The larva of another type of primitive chordate is not like that of an echinoderm but is shaped like a small tadpole. In its tail there is a hollow dorsal nerve cord and a notochord. In the forepart are the gill slits. A chordate beyond question. However, this creature in undergoing the change to the adult form discards its tail (as a tadpole does), losing in the process all its notochord and all but a tiny scrap of the nerve cord. What is left of the creature adopts a sessile habit—that is, it is fixed to the surface—and develops a thick, tough covering called a tunic, so that the creatures are referred to as *tunicates*.

Viewed as adults they seem to have nothing of the chordate about them at all, except for the fact that they do retain numerous gill slits through which water is sucked so that food might be filtered out. The filtered water is squirted through an opening in the side, and the creature is also called a "sea squirt."

So far the chordates seem to be doing very little with the notochord, but let's go back to the larva of the tunicates.

There is sometimes a tendency among animals to retain the larval form for extended periods. It may happen that the larva is better adapted to a particular set of environmental conditions than the adult so that it becomes advantageous to lay stress upon this form. Among some insects, for instance, the larvae are relatively long lived (living for a number of years on occasion) and the adult form is very short lived indeed. The adult form may have as its sole function the quick laying of eggs out of which another extended larval form might hatch. Those adults may even lack mouth parts, since in their ephemeral life span there would be no necessity for them to eat.

If the larva develops the last special ability retained by the adult form, that of reproducing itself, the adult form may be eliminated altogether and the larval form can become all there is. This is actually observed to happen among certain salamanders and the phenomenon is called *neoteny* (nee-ot'i-nee; "new stretch" G, that is, a new

creature developed through stretching out the larval stage).
This tendency also exists among the tunicates: there are
small creatures within the group in which the tail of the
larva persists throughout life.

In the long-ago Cambrian Period, then, it was possible
that some primitive tunicate underwent neoteny and the
tail portion of the animal took on a greater importance
until a new type of creature arose that was all tunicate tail.

There is a small creature existing today that could conceivably be the descendant of an early tunicate-tail. It is
about two inches long and is vaguely fishlike in appearance.
Its head end has a circular opening surrounded by bristles
which sweep a current of water into the mouth and out
the gill slits behind the head. Both head and tail come to
a relatively sharp point and the creature is therefore called
amphioxus (am'fee-ok'sus; "equally-sharp" G). Because
it also resembles a tiny lance in shape it is called a *lancelet.*

The amphioxus has a hollow dorsal nerve cord and
beneath it a notochord running the full length of the body
from end to end. This is the simplest living creature in

which an internal rod can actually act as a stiffening agent
throughout life. The amphioxus also clearly shows segmentation. The mere existence of a series of repeated structures
such as the gill slits is sign enough that segmentation is a
fundamental characteristic of all chordates, but in the
amphioxus one can see, through its semitransparency, that
its muscles are arranged in segments.

These three groups of organisms—balanoglossus, the
tunicates, and amphioxus—are so different that although
all are chordates they are nevertheless placed in three
separate *subphyla.* Balanoglossus is *Hemichordata* (hem'
ee-kawr-day'tuh; "half-cord" G); the tunicates are in

Urochordata (yoo'roh-kawr-day'tuh; "tail cord" G); and amphioxus is in *Cephalochordata* (sef'uh-loh-kawr-day' tuh; "head cord" G). Some zoologists consider Hemichordata a small phylum in its own right.

THE VERTEBRATES

By and large, the chordates as described so far are not a successful phylum. The number of species they include are few and obscure and the lives they lead are passive and backward. Yet there is great potentiality here. The notochord is the beginning of an internal bracing to which muscles may be attached. Such internal bracing is much lighter and more efficient than the external bracing of a shell. Then, too, the gill slits can be adapted to extracting oxygen as well as food from the water, so that respiration can be more efficient than in other phyla. Finally, the dorsal nerve cord proved itself in the course of some hundreds of millions of years to be capable of elaboration and development beyond that possible for any ventral nerve cord.

But all of this remains potential, rather than actual, in the small and unsuccessful trio of subphyla so far described. There remains, however, a fourth subphylum which might conceivably have developed from the ancestors of amphioxus, the one primitive group with a notochord lasting through life. It is to this fourth subphylum that man belongs, and, in fact, most of the familiar animals about us as well.

What happened was that the notochord, a continuous and unsegmented rod, took on the segmentation of the rest of the body. It was gradually replaced by a series of discs of *cartilage* ("wickerwork" L), one to each segment. Not only did this supply the new creature with a rod that was springier and more flexible, but the individual segments grew to enclose the dorsal nerve cord so as to offer that key portion of the organism important protection from shocks and buffets. Strips of cartilage also lined the gill slits, stiffening them and forming the *gill arches*.

The individual discs into which the notochord had been converted are called *vertebrae* (vur'ti-bree; singular, *vertebra*), for reasons I will explain later. Creatures with such vertebrae make up all the rest of the Chordata and are included in the fourth and last of its subphyla, *Vertebrata* (vur'ti-bray'tuh). It is this subphylum which includes the "vertebrates" of Lamarck.

All vertebrates possess the hollow dorsal nerve cord characteristic of chordates, enclosed by the vertebrae. (We ourselves do, which makes man a member of Chordata and of Vertebrata.) However, the vertebrates, having gained the vertebrae, have lost the notochord. Should this give them the status of a separate phylum? Perhaps it might, if the notochord had really been lost, but it hasn't. To qualify as a chordate, it is only necessary that an organism possess a notochord at some time in its life, as the tunicate does in its larval stage.

Now a vertebrate such as man has no larval form in the usual sense, but he does develop in stages from the original fertilized ovum. From the time of the fertilization of the ovum to the time of actual birth, a period of some nine months elapses. During that time the human being is systematically developing within the mother's body as an *embryo* (em'bree-oh; "inner swelling" G). The human embryo has not been studied as well as have embryos of other creatures more available for experimentation and dissection, but the main line of development is clear. During the third week of development, for instance, there is an unmistakable notochord present in the human embryo. As the days pass, the tissues about it segment, and form blocks which absorb and replace the notochord, forming vertebrae instead. However, the notochord was there for a while and man (as well as all other members of Vertebrata) is therefore a full-fledged chordate.

25-DAY EMBRYO

The subphylum Vertebrata is divided into a total of eight narrower classifications called *classes,* and these are in turn grouped by fours into two *superclasses.* If we describe the nature of these classes briefly, we shall continue to get glimpses of the evolution of man; of the manner in which little by little structural refinements were added and extended until we ourselves existed on the face of the earth.

The first of the two superclasses of Vertebrata is *Pisces* (pis'eez; "fish" L) and includes all the vertebrates that are primarily water-dwelling creatures. The most primitive of the four classes of Pisces must have included the amphioxus-like creatures that first developed cartilaginous nerve-enclosing vertebrae. Like amphioxus, they retained a circular mouth opening without the type of jaw that can open and close. For that reason, the class is called *Agnatha* (ag'nuh-thuh; "no jaws" G).

The original agnaths must have been harmless filter-feeders like the modern amphioxus, but the few agnaths that remain in existence today have learned new tricks. The best known is the lamprey, its circular mouth equipped with hard little rasps, which attaches itself to a fish and plays the role of vampire.

The original agnaths, half a billion years ago, went on to develop a second improvement. In a number of phyla, as I have explained, protection was introduced by means of a hard outer shell and the agnaths fell back on this same device. One group of the creatures developed a shell over the head and forebody and are therefore termed the *ostracoderms* (os'truh-koh-durms; "shell-skin" G).

These agnath shells were not just another set of shells, however. They represented a vital new departure. Instead of being composed of calcium carbonate, as were those of the mollusks, they were composed of calcium phosphate. The calcium phosphate structures produced by ostracoderms is called "bone" and it is a substance unique to the vertebrates. It is found nowhere else in the domain of life. The advantage of bone as compared with shells of other materials lay in its uncommon toughness and strength. A section of bone that is one square inch in cross-section has a tensile strength of almost six tons; it would take that much force to pull it in half.

The next step, once the vertebrates were so efficiently defended, was to furnish equipment for possible aggression. The first of the gill arches, the one nearest the circular opening that served the agnaths as a mouth, gradually bent in two and, hinged at the middle, became a primitive jaw. This was enough of a change to warrant placing the newly jawed creatures into a separate class. Since they retained the bony foreshell of the ostracoderms, the new class is called *Placodermi* (plak'oh-durm'eye; "plated-skins" G). The Placodermi could, with the development of teeth, seize food, tear it into bits, and swallow it. The gill slits lost their food-straining properties and began to specialize for breathing only.

The placoderms are now extinct, the only class of Vertebrata to have no living representatives at all. They were successful in their time, but they gave rise to new classes of Pisces that replaced them. The new classes did away, by and large, with the external armor, and relied on maneuverability and speed rather than passive defense. In evolutionary progress as in human warfare, this is often a good move. The ostracoderms are also extinct, but some of the unarmored varieties of Agnatha—just a few species —are still in existence today.

From the placoderms were developed the remaining two classes of Pisces. In both classes the external armor was abandoned as such. Some was discarded altogether and the rest was overgrown with skin so that it became an internal protection enclosing the fore end of the nerve cord, which had grown and specialized into a primitive brain.

These two classes further developed movable, paired fins. The agnaths and placoderms possessed fins (sometimes a considerable number of them) along the midline of the body. They served as balancing organs and kept them right side up when swimming. They were stiffened by cartilaginous fin-rays.

The new classes converted these into two paired fins, located on either side of the midline, one pair just behind the head and the other just before the tail. These were stiffened not only by cartilaginous rays but by internal stiffening rods extending down from the vertebrae. To these supports muscles could be attached which could move the fins and convert them from passive balancing devices to oars that could help bring about rapid turns and all sorts of agile maneuvers. From the individual vertebrae, curved extensions were formed which stiffened the sides of the creature. The internal bracing that had begun with the simple rod of the notochord had thus become a complicated system including a jointed vertebral column with a brain-enclosing skull at one end, a set of ribs enclosing the body, and limb supports.

The two new classes differed in one important respect in their treatment of bone. One class retrogressed, abandoning bone altogether and forming a system of bracing entirely out of cartilage. This is the class *Chondrichthyes* (kon-drik'thee-eez; "cartilaginous fish" G), represented today by the various sharks. The second class, which makes up the last of the Pisces, retained the bone, shifting it inward. They even converted the cartilage of the vertebral column and its extensions into bone. This is the class *Osteichthyes* (os'tee-ik'thee-eez; "bony fish" G) and it is to this class that all the familiar fish of today belong.

The bony fish came into prominence in the Devonian Period (named for a region in southern England where rocks of that period were first studied) about 400,000,000 years ago. It took a hundred million years of chordate evolution to do it, but the phylum finally made its mark. The bony fish dominated the ocean and proliferated into a large number of species. The period is sometimes called the "Age of Fishes" in consequence. As far as the ocean is concerned, in fact, the Age of Fishes has never ended, for the bony fish still dominate the ocean today.

Most of the varieties of bony fish kept their paired fins largely in the form of thin fringes of ray-supported tissue. The bony supports were small and weak, only as strong, in fact, as was necessary to maneuver the fins as oars. It was not from these, the most successful fish, that man descended but from poor relations of these ray-finned fish.

Among these poor relations, the fleshy part of the fin,

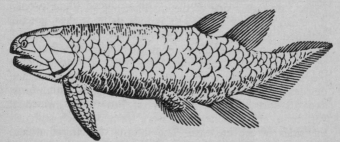

together with its bone and muscle, was expanded so that each of the four fins seemed to form a stubby extension of the body, a kind of fleshy lobe with but a thin extension of ray-supported fringe. These are the "lobe-finned fishes," or *crossopterygii* (kros-op'tuh-rij'ee-eye; "fringe-fins" G).

The lobe-finned fishes sacrificed swimming agility by this sort of fin formation and were far less successful than the other groups of bony fish. They were thought to have become extinct about 70,000,000 years ago, but in 1939 a living lobe-finned fish was netted in the waters off South Africa and several have been caught since World War II. This only means that a tiny remnant have managed to hang on through the ages. The lobe-finned fishes gained an advantage, however, in shallow, swampy water, where their stubby fins, poor at swimming, showed up well as supports. They could hobble out of a drying stretch of water into a deeper pool.

The limbs strengthened further, and with other adaptations, involving lungs and heart, developed into forms of life capable of living on dry land, first for extended periods and finally permanently. Thus, although the descendants of the lobe-finned fishes may have perished—or largely perished from the waters—other descendants invaded the dry land successfully and gave rise to the second superclass of Vertebrata, the superclass to which man belongs.

The members of this second superclass could no longer rely on water buoyancy to help their movements. They had to struggle against the full force of gravity. Their limbs became larger and stronger. Those remained four in number for the most part, though some members of the superclass, such as certain flightless birds, reduced them to only two functional ones, and others, such as the snakes, got rid of them altogether. Nevertheless these are exceptional and no land vertebrate ever developed a fifth limb. For this reason, the entire superclass is termed *Tetrapoda* (te-tra'poh-duh; "four-footed" G). We make use of about the same term when we speak of the animals with which we are most familiarly acquainted as quadrupeds, which is Latin for "four-footed."

The tetrapods are divided into four classes, of which the first was developed from the crossopterygii about

300,000,000 years ago. This includes creatures that are not yet entirely emancipated from water. That is, they lay eggs in water and the larval forms that hatch out are rather fishlike, with finny tails and gills, but with no legs. Eventually these larvae undergo a radical change, substitute lungs for gills and legs for a tail. Their adult life they spend on land (but near water, usually). Since their life is spent both in water and on land, they make up the *Amphibia* (am-fib'ee-uh; "double life" G). The frogs, toads, and salamanders are the best known of still existing amphibia.

Eventually certain descendants of the amphibians developed eggs that could be laid on land and thus freed themselves of the waters. These were a second class of tetrapods, *Reptilia* (rep-til'ee-uh; "to creep" L, from the fact that the most spectacular of modern-day reptiles, the snakes, progress in that fashion). The amphibians and reptiles, though each witnessed a time when their members were the dominant forms of life on land, are today rather unsuccessful classes. They were replaced and supplanted by certain reptilian descendants that reached new heights of efficiency.

The reptiles and amphibians, together with all members of the superclass Pisces, and, indeed, with all members of all the phyla outside Chordata, are cold-blooded—their internal temperature tends to be that of the external environment. Descended from the reptiles, however, are two classes of animals which independently developed, about 150,000,000 years ago, the faculty of being warm-blooded; that is, of having an internal temperature kept constant at a point generally higher than that of the external environment. This was something new and unique, with advantages I shall discuss later in the book.

The first of these classes to come into existence was that of *Mammalia* (ma-may'lee-uh; "breast" L, so called because the members of the class possess milk-producing organs for feeding their young). The fourth and last class was that of *Aves* (ay'veez; "birds" L), a class commonly referred to as "birds." The easiest way of distinguishing between these two warm-blooded classes lies in their method of insulating their bodies against the excessive loss of body heat. The birds use feathers for the purpose and

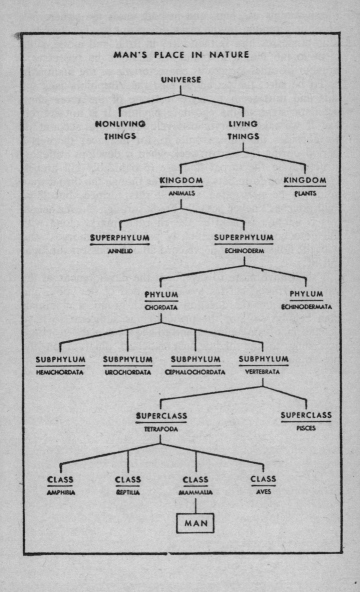

MAN'S PLACE IN NATURE

UNIVERSE

NONLIVING THINGS | LIVING THINGS

KINGDOM
ANIMALS

KINGDOM
PLANTS

SUPERPHYLUM
ANNELID

SUPERPHYLUM
ECHINODERM

PHYLUM
CHORDATA

PHYLUM
ECHINODERMATA

SUBPHYLUM
HEMICHORDATA

SUBPHYLUM
UROCHORDATA

SUBPHYLUM
CEPHALOCHORDATA

SUBPHYLUM
VERTEBRATA

SUPERCLASS
TETRAPODA

SUPERCLASS
PISCES

CLASS
AMPHIBIA

CLASS
REPTILIA

CLASS
MAMMALIA

CLASS
AVES

MAN

the mammals use hair, and in both cases the structure is absolutely unique to the class possessing it. The fact that man possesses hair is therefore, in itself and alone, sufficient to tab him a mammal. In addition he possesses a number of other structures characteristic of the mammals.

To be sure, the tetrapods past the Amphibia lack gill slits and thus seem to have lost one of the three characteristic signs of the chordate phylum. It is not entirely lost if we take embryonic development into account. If we consider the human embryo, for instance: there is a period during the fourth week when it develops stiffening structures in the throat that are recognizably gill arches. Hollows even form between them as though the throat were going to be perforated and gill slits formed, but these "gill pouches" never actually break through. Nevertheless, this is enough to lend us the chordate mark, along with the notochord we momentarily possess as embryos, and with the hollow dorsal nerve cord all of us possess throughout life.

I shall have more to say about the development of the various tetrapod classes toward the end of the book, and I shall consider the question of which group of mammals man belongs to. For the present, it is sufficient to have defined man as a mammal and to show his place in nature according to the highly schematicized diagram on the preceding page.

2

OUR HEAD AND TORSO

THE VERTEBRAL COLUMN

That which makes man most clearly a vertebrate (even to the untutored eye), which most firmly knits him to other members of the subphylum, which most markedly sets him off from creatures outside the subphylum is his internal framework of bone. The initial chapter on man's place in nature would thus seem to lead us to this bony bracing as the logical place to start considering the human body.

The bones of our body (and, indeed, of any vertebrate body) make up the *skeleton* ("dried up" G, because a skeleton by itself has a resemblance to a completely dried up human being, a shriveled mummy with even the skin removed). In fact, the skeleton, which is the framework about which the "soft tissues" are molded, gives a clear indication of what the human form in its entirety is. The same can be said for other members of our subphylum and it is by finding fossilized skeletal remains of long-dead creatures that paleontologists are able to reconstruct the appearance of animals that lived hundreds of millions of years ago.

In the case of man, the grisly similarity of skeleton to body, the wide grin, the slitted ribs, the overlong fingers,

at once showing the man, yet distorting him, lends the skeleton a frightening aspect to children and to unsophisticated adults. We ourselves shall, of course, proceed to view the skeleton unemotionally, and even statistically.

The skeleton makes up about 18 per cent of the weight of the human body; some 25 pounds, in other words; and is made up of a little more than 200 individual bones. Of these, the oldest, evolutionarily speaking, is the series of bones running the length of the back to form the central axis of the body—the bones that mark the original site of the notochord. The common name for this line of bones is *backbone,* which is descriptive enough but implies the existence of a single bone, whereas a family of more than two dozen bones actually is involved.

Each of these bones is irregularly shaped, with several projections pointed enough to resemble spiny outgrowths. If you bend your back forward, you will be able to feel a line of these projections. If you observe someone else's curved back, they will show up as a line of knobs. Since this is the most clearly visible characteristic of the backbone in living man, it is not surprising that another common name for it is the *spinal column* (that is, a column of spiny bones). Sometimes, this is shortened to *spine,* which saves breath but again mistakenly implies the existence of a single bone.

If the spinal column were indeed a single bone, the back would be rigid and unbending; just as the thigh, which is built about a long single bone, happens to be. It is because the column is a column and one built of separate bones that the trunk can bend forward, backward, to either side, or even go through small circular movements. Furthermore, it does not bend sharply at some individual point, as the arm bends at the elbow, but only slightly at each of a number of points, forming a smooth concavity or convexity. In this way it retains something of the strength of a rigid bone and something of the flexibility of truly jointed bones. It is a very successful compromise.

The property of the spinal column which allows turning and bending in various directions gives it its most formal name, the *vertebral column* (vur'ti-brul; "to turn" L).* The individual bones of the column are therefore *vertebrae,* and that is how members of our subphylum come to be called *vertebrates.*

In the various sea creatures of the superclass Pisces the spinal column forms a straight line, running horizontally when the creature is in the usual swimming position. The individual vertebrae are very much alike.

In land animals such a simple arrangement is not practical. Whereas in sea creatures the body is supported at every point by water buoyancy, in tetrapods the body is supported on four legs, of which one pair are at the front end of the vertebral column and the other pair at the back end. From the stretch of column between forelegs and hindlegs are suspended various organs, pulled downward naturally by the force of gravity. If, under these conditions, the column were straight, the suspension of weight would inevitably lead to a downward curve—a "swayback." To prevent that, the spinal column in tetrapods is arched so that each vertebra rests partly on the bone immediately before or behind it. The pull of the suspended organs is transferred down the line of vertebrae to the forelegs and hindlegs.

In the tetrapod, moreover, the individual vertebrae be-

* I must warn you that the derivation given is not necessarily a direct translation of the scientific term. Thus "vertebral" is from the Latin *vertebralis,* meaning "of the spine," but that in turn comes from the Latin *vertere,* meaning "to turn." In my derivations I shall, when necessary, skip the intermediate stages and get back to what seems to me to be the significant root.

gin to show differentiation; that is, they show differences
in form that suit them to differences in function. This
might also be termed specialization, since the individual
bones are modified to suit special functions.

After all, when a fish must turn, a flip of the tail will
turn the whole body easily enough, suspended as it is in
fluid. On land, a vertebrate is not so fortunate. To turn
its body, it must initiate a series of complicated leg movements. When the turning is only for the purpose of bringing the sense organs (concentrated in the head) to bear in
a new direction, it would be very convenient if the head
were turned without involving the limbs.

The ability to do this was indeed developed, and for
the purpose a narrowed neck region was evolved. This
region formed about a section of the vertebral column
that was specialized in form to allow a greater freedom of
bending. The line of vertebrae in the neck region also
curved upward to give the head greater height for vision
over a greater area.

When a land vertebrate returns uncompromisingly to
the sea, as the whales and porpoises have, these modifications for land life are done away with. In the whale and
its relatives, the vertebral column is once again quite
straight and the neck region has all but disappeared.

In man, there is a new change. When born,
we have the two-curved vertebral column of
the typical land vertebrate, a downward curve
at the neck and an upward curve in the back.
To get around, infants crawl quite handily in
the usual tetrapod fashion. In the second year
of life, however, the infant rises upon his hind
legs and finds it increasingly comfortable and
natural to remain so. In order to make that
possible, the human vertebral column bends
backward in the region of the hip, forming a
new curve that is concave toward the back.
The human vertebral column, although still
perfectly straight when viewed from the back,
displays a series of rather graceful curves, a
kind of double S-shape, when viewed from the
side.

Despite the fact that the human body seems

to be precariously tipped upright, the curves of the vertebral column make the position a relatively easy one to maintain and lend us a springy balance. Other animals that are capable of rising to their hind legs, such as bears and gorillas, lack this vertebral curve in the hip regions and therefore cannot maintain the upright position for long. The gorilla, consequently, rarely is truly erect, but hunches forward for lack of the vertebral curve and supports itself partly on the knuckles of its hands.

To be sure, there are two-legged creatures, such as the kangaroos and birds, which retain an essentially horizontal vertebral column. They manage to retain balance by the development of a tail set far enough back so as to act as a counterweight for the forepart of the body. (The penguin is an exception, with a rather humorously manlike waddle.)

This tipping of the vertebral column in man raises certain difficulties in comparing positions in man with those in other animals. Anatomists use *dorsal* ("back" L) to mean "toward the back." Although this is really toward the back as far as the human being is concerned, it is toward the top in most other animals. (We run into the same thing in using expressions of Anglo-Saxon origin. We speak of riding "horseback" but the horse's back is not at the back of the animal but at the top.)

Again, there is *ventral* ("belly" L), which means "toward the belly"; that is, frontward in man, but downward in most other animals. The expressions *anterior* ("farther forward" L) and *posterior* ("farther backward" L), which mean toward the head and tail, respectively, when referring to positions in most animals, mean toward the belly and back where man is concerned.

Perhaps the safest way to avoid confusion is to forget about up, down, forward, and backward altogether and define the directions in terms of parts of the body. In most vertebrates dorsal is toward the vertebral column, ventral is toward the belly, anterior is toward the head, and posterior is toward the tail. In man, *superior* is toward the head, *inferior* toward the feet.

THE VERTEBRAE AND RIBS

Since the notochord originally lay ventral to the nerve cord, the main portion of the typical vertebra is still ventral. Here there is a solid cylinder of bone called the *centrum,* or *body,* of the vertebra. From it arises dorsally an arch of bone, enclosing a roughly circular space. The enclosed space is the *neural canal* (nyoo'rul; "nerve" G) and the bony arch that forms it is the *neural arch.* As you can guess from the derivation, the dorsal nerve cord runs through that ring. In fact, it runs through a whole series of rings, formed by successive vertebrae.

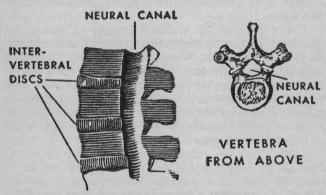

NEURAL CANAL

INTER- VERTEBRAL DISCS

NEURAL CANAL

VERTEBRA FROM ABOVE

The neural arch possesses three projections, or processes. One emerges dorsally (sometimes angled a bit downward) and it is the one you feel when you run your finger down the vertebral column. These being the "spines" you feel, it is called the *spinous process.* The other two emerge to either side and are the *transverse processes.* The word, transverse ("move across" L) implies a direction that is neither forward nor backward but sideways.

If you have ever nibbled at the neck of a chicken, you know how irregular and sharp these processes can be. These irregularities are not useless ornaments, but serve an important purpose, for it is to these processes that various muscles are attached and to which other bones may join.

The human vertebrae are divided into differentiated groups. The seven most superior of them, for instance, are the *cervical vertebrae* (sur'vi-kul; "neck" L), which are, as the name implies, the vertebrae of the neck. The fact that there are seven of these cervical vertebrae is typical of the mammals. With the exception of a couple of species of sloths, all mammals, regardless of the length of the neck, possess just seven, no more and no less. In the case of the whale, which has no neck, the seven vertebrae are flattened to negligible size, yet there remain seven of those flattened discs. As for the giraffe, the full length of its neck contains but seven cervical vertebrae, though these have lengthened out until they look more like bones of the limbs than like vertebrae.

Birds, on the other hand, though not as rigid in the number of cervical vertebrae as mammals are, usually possess about twice the mammalian number. And so (a fact dear to the heart of "Believe It or Not" columnists) the sparrow has more bones in its neck than the giraffe does. It follows also that birds such as the swan or flamingo have much greater range and grace of movement in their long necks than the giraffe has. In fact, while to say a girl's neck is "swanlike" can be used as a great compliment, to say it is "giraffe-like" would be to invite disaster. (To point out that, anatomically, a girl's neck *is* giraffe-like and not swanlike would probably do no good, either. In fact I am sure it would just make matters worse.)

In man the first cervical vertebra has a specially modified shape to allow it to be connected with the bony structure of the head. It has no real centrum but is all neural arch. What's more, it is a large neural arch, for the nerve cord is at this point widening to become the brain.

When you nod your head, bending takes place chiefly between the skull and the first vertebra. Because in man the roughly globular structure of the skull rests upon the neural arch of this first vertebra, like the globe of the earth resting upon the shoulders of the giant Atlas in the Greek myth, this vertebra is called the *atlas*.

When you shake your head from side to side, the atlas moves with the skull, the motion taking place along the division between the first and second cervical vertebrae.

The second vertebra possesses a special process, jutting upward. Over it, the atlas fits neatly, and it is about that anterior process as an axis that the head makes the gesture of *No*. The second vertebra is therefore commonly called the *axis*.

The spinous processes of the cervical vertebrae are rather sharp and are slightly forked just at the end.

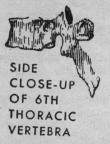

SIDE CLOSE-UP OF 6TH THORACIC VERTEBRA

Inferior to the cervical vertebrae are twelve *thoracic vertebrae* (thaw-ras'ik; "chest" L) which run down the length of the chest. (These are sometimes called the *dorsal vertebrae,* but this is a poor alternative, for all the vertebrae are dorsal, really.)

The thoracic vertebrae have somewhat longer than average transverse processes for to them are attached *ribs*.* There is one rib attached to each transverse process of each thoracic vertebra, making twelve pairs of ribs, or twenty-four in all. Where the rib meets the vertebra, it extends two processes, one adjoining the transverse process of the vertebrae and the other adjoining the centrum itself. Each rib curves in a downward semicircle, so that taken together a pair encloses the chest. Most of the pairs come together ventrally and join a flat bone extending down the midline of the front of the chest. This is the *sternum* ("chest" G), or the *breastbone*.

The first pair of ribs are comparatively short, but each of the next six pairs are successively longer. All seven pairs join the sternum directly and are sometimes called the "true ribs" in consequence. Pairs eight, nine, and ten are the "false ribs." They do not join the sternum directly but converge and join the seventh rib at a point before the sternum is reached. There is thus a sharp upward notch in the bony structure in front of your chest, which you can easily feel if you follow the line of your lower ribs. The eleventh and twelfth pairs of ribs do not complete their

---

* One of the anatomical terms of Anglo-Saxon derivation, which I shall henceforth indicate, when it seems helpful by a simple AS in parentheses.

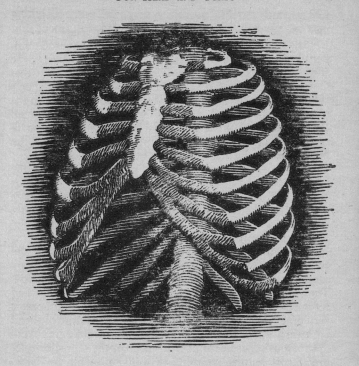

curve, but come to an end in midchest, so to speak. These are the "floating ribs." The ribs and breastbone, taken all together, can be referred to as the "rib cage."

The fact that the ribs occur in pairs is an evidence of our bilateral symmetry. For a bone to occur singly it must lie along the midline of the body. Examples are the various vertebrae and the sternum. Any bone that lies to one side of the midline possesses a sister bone, mirror image in shape, that lies on the other.

The vertebrae and ribs, taken together, are the most obvious indication of the fundamental segmentation of the human body. This sort of indication is much more dramatic in certain other vertebrates. Reptiles generally have ribs attached to every vertebra except those in the tail. A large python, with a couple of hundred vertebrae running the sinuous length of its body, has a couple of hundred pairs of ribs, and its skeleton gives the unmistakable impression of being a monstrous centipede.

Incidentally, it might be well to emphasize that the number of ribs in men and women are identical. Because of the story in the book of Genesis (2:21–22) to the effect that God made Eve out of a rib he withdrew from Adam's side, it is sometimes suggested that one rib is missing in the male. Not so. In fact, although anatomists can easily tell the skeleton of a male from that of a female because of the difference in the shape and proportion of some of the bones, neither sex has a bone that the other lacks.

But let's return to the vertebral column. Inferior to the thoracic vertebrae are five *lumbar vertebrae* (lum'bur; "loin" L). These lack ribs and form the narrow "waistline" of the skeleton. It is also because they lack ribs that some women can delight in a narrow waistline that accentuates the broader areas above and below.

The lumbar vertebrae are the largest and heaviest of the column, because, thanks to man's upright posture, they support the weight of the entire upper half of the body. The spinous processes of these vertebrae are stubbier and set farther apart than are those of the cervical and thoracic vertebrae. This makes it possible for a person to bend backward at the waist through a considerable arc. Were the spinous processes similar to those of the thoracic vertebrae, their mutual interference would prohibit the backward bend.

Inferior to the lumbar vertebrae are five *sacral vertebrae* (say'krul; "sacred" L) which differ in several ways from all those I have described previously. In the first place, they are not named after their position in the body, in contrast to the previous groups of vertebrae which are named after neck, chest and loins. Actually, no one is certain why they are named as they are. The easiest assumption is that they had some special significance in Roman religious rites, but that is just an easy assumption, and it is not necessarily the truth.

Then, too, the five sacral vertebrae are separate in young children, but with age, the spaces between become ossified and they fuse into a single bone called the *sacrum* (say'krum): In adulthood all that remains to indicate the original separation are four transverse lines where the

joining had taken place and four
pairs of holes, one of each pair on
either side of the midline. These
holes are formed when the processes of the adjoining sacral vertebrae fuse together. Holes such
as these through a bone or through
any otherwise continuous structure in the body are called *foramina* (faw-ram′i-nuh; "to
pierce" L) the singular is foramen
(faw-ray′men).

The case of the sacrum shows
that it isn't possible to be too hard
and fast in describing the body.
The body is not a machine turned
out by some known and unvarying process, but has its
individual idiosyncrasies. It is easy to say, for instance,
that there are 33 bones all told in the human vertebral
column, but does the sacrum count as one bone or as five?
It is clearly five in the infant, and as clearly one in the
adult. If it is counted as one then there are only 29 bones
in the human vertebral column. There are other cases
where bones may or may not be fused in different individuals, so that in the end one ought only to speak approximately of the number of bones in the human body, as I did
at the beginning of the chapter when I said there were a
little more than 200. Sometimes the number is stated precisely as 206, but that is not always so.

The sacrum forms a strong bone to which the bones of
the hips and hind legs can be securely attached. It is proportionately larger and stronger in man than in other mammals, because man's upright structure places considerable
weight upon it. On the other hand, mammals that have
adapted themselves to the sea to the point where they
have no hind legs at all (as in the case of whales and sea
cows) have no sacrum, either—merely a line of lumbar
vertebrae running down to the tail. Since the sacrum is
intimately connected with the bones of the hip and a
woman's hips are proportionately wider than are those of
a man, the sacrum in the female is likewise proportionately

wider. Here, then, is one way in which an anatomist can tell the skeleton of a woman from that of a man.

The final, most inferior group of vertebrae are conspicuous indeed in most mammals, since they are the *caudal vertebrae* (kaw-dul; "tail" L). They are numerous in mammals with long tails. It might be thought that a human being, who clearly does not have a tail, would as clearly lack caudal vertebrae. But this is not so.

Below the human sacrum are four small vertebrae, each smaller than the one above, and each without a neural arch. These make up the remnant of what might have been a tail. (Some individuals may have five—another reason for not being too dogmatic about the number of bones in the human body.) Together these final vertebrae make up the *coccyx* (kok'siks; "cuckoo" G, so named because the whole seems to resemble the beak of a cuckoo in shape). As a result, the individual vertebrae are sometimes called *coccygeal vertebrae* (kok-sij'ee-ul).

If there is any doubt that the coccyx represents a tail and not something else entirely, the answer lies in the study of the developing human embryo. In the early stages a small but distinct tail region is formed. By the eighth week of development it is gone, but its evanescent existence would seem to make it clear that man descended from some creature with a tail, and that he still carries about with him, hidden below the skin, a last evidence of it. (It is interesting, by the way, that the gorilla seems to have moved farther from a presumably tailed ancestor than we have: the gorilla is down to three caudal vertebrae as opposed to our four.)

The vertebral column is not made up of bone alone. It also contains cartilage, the structure out of which the column of the first vertebrates was formed. A baby is born with its skeleton still largely in the cartilaginous stage, and ossification proceeds throughout all the years till maturity. As an example, those portions of the ribs adjacent to the sternum are actually bars of cartilage, called the *costal cartilages* ("rib" L). These costal cartilages are extensive in the infant, but become much shorter in length in the adult.

In the vertebral column particularly, the elasticity and flexibility of cartilage plays an important part. Between the individual vertebrae are stubby cylindrical discs of fiber and cartilage containing a gelatinous material in the center, which may be the last remnant of the original notochord. These discs are consequently spongy and compressible, and make possible the smooth bending of the vertebrae. They also act as shock absorbers, so that the column can take the sudden pressure changes that result, let us say, from jumping off a six-foot height or lifting a hundred pounds. In old age the discs lose the gelatinous center and become all cartilage. This is responsible for much of the characteristic stiffness of age.

The intervertebral discs have also gained a certain notoriety that stems, really, from the flaws inherent in our upright posture. The upright posture of man is extremely useful from the standpoint of freeing his arms and hands for purposes other than locomotion. It also gives him added height so that he might make more efficient use of his head-centered senses. Nevertheless, it remains a monstrous perversion of tetrapod structure.

For some hundreds of millions of years, the structure of land vertebrates has been designed to fit an internal skeleton consisting of a more or less horizontal (though arched) vertebral column set firmly on four supports. Over the space of a single million of years or less, various human and prehuman species have tipped the whole structure on end. While the ingenuity of the adjustment to such a change is impressive, it must be admitted that the vertebral column is not completely adjusted to the new situation.

It is possible, as the result of a momentary over-great stress, to cause one of the discs to protrude slightly from between the vertebrae. This is most likely to happen in the lumbar region where, thanks to upright posture, stresses can be almost unbearably concentrated. Such a "slipped disc" will, naturally, pinch the nearby nerves and can result in excruciating pain—the price we still pay (among others) for getting up on our hind legs some hundreds of thousands of years ago.

THE SKULL

The superior end of the spinal column is connected to the *skull* (AS), which makes up the bony framework of

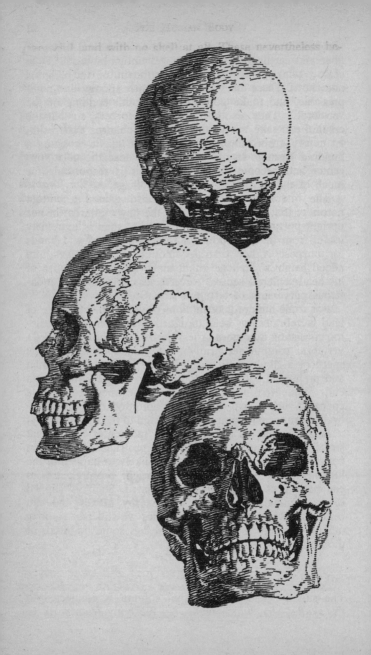

the head and face. The main portion of the skull is a nearly smooth, roughly ovoid structure called the *cranium* ("skull" L). It almost entirely encloses the brain, which is thus the only organ in the body to have such a close-fitting coating of bone. In fact, the brain can almost be viewed as enclosed in a shell. It should therefore not be surprising that the word "skull" may be cognate to "shell," both coming from the Scandinavian word *skel,* meaning a seashell. A consonantal change in one direction and a vowel change in the other ended by giving us two words.

At the base of the skull is a hole called the *foramen magnum* ("large hole" L), which fits over the enlarged neural arch of the atlas, that is, the first vertebra. Knobs at the bottom of the skull, on either side of the foramen magnum, fit neatly into depressions in the atlas. Such a bony knob is called a *condyle* (kon'dil; "knuckle joint" G). Through the foramen magnum the thickening nerve cord makes its way, and within the cranium expands to form man's large brain. In a way, the skull may be looked upon as forming a gigantic dead-ended neural arch.

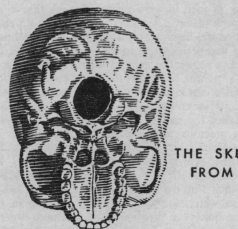

THE SKULL FROM BELOW

This development of a specialized and quite intricate bony structure about the anterior end of the original nerve cord is the end product of a process that began quite early in evolutionary history. Once a multicellular organism with bilateral symmetry was developed (long, long before our fossil record begins), it became possible to have a preferred direction of movement. At one end of the plane of symmetry would be the head and at the other the tail, the head being defined as the end which lay in the direction in which the animal moved. This meant that it was the head that was always breaking new ground, so to speak, and advancing into a new and untried environment. Specialized organs for detecting changes in the environment were most useful if they were located in the head. In order to receive and correlate the impressions received by such "sense organs," the forward end of the nerve cord had a tendency to grow more complex.

This tendency to enrich the head with sense organs and to swell the forward end of the nerve cord is called *cephalization* ("head" G) and it is to be found in many phyla.

Chordata, by shifting the nerve cord to the dorsal position, seemed in a sense to have to begin all over. At least amphioxus (a chordate but not a vertebrate) is quite uncephalized. It has no advanced sense organs, no nerve-cord swelling; in fact, its very name indicates that it has no head to speak of but is "equally pointed" at both ends.

Cephalization did begin with the subphylum Vertebrata, however, and was here carried to the greatest extreme the realm of life can show. The agnaths, which first developed cartilaginous vertebrae to protect the nerve cord generally, also developed a cartilaginous box to enclose and protect the enlarged anterior end of it. In addition, the agnaths and their descendants, the placoderms, developed a bony shell to protect the precious and specialized head region.

Oddly enough, this bony shell, though it seemed to disappear with the extinction of the placoderms and the armored agnaths, has left its mark upon all their unshelled descendants, including ourselves. The proof of that lies in the manner in which the various bones develop in the embryo. Most bones of the body are produced by the ossification of previous cartilaginous structures, which, in this fashion, act as models for the final product. These are "cartilage replacement bones," and the vertebrae, ribs, and sternum are all examples. The human skull, however, does not develop from such a cartilaginous model. Instead, the bones that make it up begin to form beneath the skin, as though harking back to a long-past time when such bone formed on the outside rather than the inside of the body. The skull is apparently a relic of the external armor of the placoderms which has been narrowed in function. Instead of enclosing the head and fore regions generally, it has been drawn within and set to work as a tight enclosure for the brain and as a specific protection for the most specialized and most vulnerable of the sense organs, the eyes and ears.*

In the lower vertebrates the skull tends to be quite complicated in structure. The evolutionary tendency has been in the direction of greater simplicity and fewer individual bones. Fish have more than 100 bones in the skull, some reptiles as many as 70, and primitive mammals as many as 40. In contrast to this, the human skull contains only 23 bones, and of those only 8 suffice to make up the cranium. There is sense to this because a bony structure, designed for protection, is naturally weakest at the joints; the fewer of the joints, the stronger the structure.

Of the eight bones of the cranium, the most prominent is the *frontal bone*, a single bone making up the forehead and the forward half of the top of the skull. The frontal bone reaches down to the bony circle enclosing the eye, which is called the *orbit* ("circle" L), and to the top of the nose. Just above each eye there is a low ridge stretching across the frontal bone, which may originally have served as further protection for the eye. It is very pronounced in the apes and quite pronounced in early species of man. The ridge is still present in the adult male, but is virtually absent in children and in the adult female (which is why the forehead of a woman is so attractively smooth).

Behind the frontal bone, forming the frame of the rest of the top of the skull, is a pair of bones that join at the midline of the top of the skull. These are the *parietal bones* (puh-ry'i-tul; "wall" L), which do indeed seem to be the walls of the brain. Still farther behind is a single bone forming the undersurface of the rear of the skull. This is

---

* I shall not deal in this book with the brain, nerves, and sense organs except in passing. This is not because the subject is unimportant, but rather because it is of most particular importance. I am planning a companion volume to deal with the nervous system in detail.

the *occipital bone* (ok-sip'i-tul; "away from the head" L).
This rearmost portion of the skull is sometimes called,
even in common speech, the *occiput* (ok'si-put).

On either side of the skull, beneath the parietals, are the
two *temporal bones*. These are located in the portion of
the head usually referred to as the temples. Both "temporal" and "temple" come from a Latin word meaning
"time." There are several theories as to the connection between time and this side portion of the skull. None of
these sound completely convincing, but the most interesting one is that since hair tends to go gray at the temples
first, that portion of the head most clearly marks the passing of time.

The six bones so far mentioned (frontal, occipital, two
parietals, and two temporals) make up the main structure
of the cranium. Two more bones remain, which are less
evident because they are below the cranium and in life are
hidden from us by the eyes. These are the *sphenoid bone*
(sfee'noid; "wedge-shaped" G) and the *ethmoid bone*
("sieve-like" G).*

One might expect that with continued evolution, the
number of bones in the skull might diminish further, and
so they might. There seems nevertheless to be a limit to
how far the decrease in number can go.

The bones of the cranium of the newborn child are not
joined. There are six sizable gaps, yet unossified, in the
skull at birth. These are called *fontanelles* ("little fountain" L) because the pulse of blood vessels can be felt
under the skin in those areas, so that doctors were re-
minded of the spurting of a fountain. In ordinary language
they are the "soft spots," and the largest of these is at the
top of the skull. Any parent is tenderly aware of the presence of that one, particularly in a first child.

The presence of such a loose structure of the cranium in
the newborn is essential to normal birth. The skull is the
largest portion of the fetus, and if that passes through the

---

* The first giving of names to the parts of the body (and, indeed,
scientific naming in general) was and is a tedious and difficult process;
anatomists, both ancient and modern, do their best, however. Colorful
images are found, as when the ethmoid bone is compared with a sieve
because it contains a number of foramina. Or the particular shape
of the bone is used as inspiration for the name, as in the wedge shape
of the sphenoid.

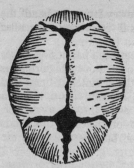

birth canal the rest of the body can follow without trouble. For the skull to get through, a certain amount of distortion is often necessary and it is these spaces between the bone that make distortion possible.

After birth, ossification proceeds and by the second year even the largest fontanelle is closed. Yet complete ossification does not take place till maturity, a fact that is also vital, since only with the joints relatively open can the brain case increase in size and allow for the growth of the brain.

Nevertheless, once birth and growth are finished, the skull, having eaten its cake, metaphorically speaking, proceeds to have it also, since the various bones seal tightly together. The boundaries formed are ragged uneven lines, as though each bone tried to grow as far into its neighbor as possible and the battle were exactly even, with one having the upper hand at this point and the other at the next point. Such an intricate, meandering joint is a *suture* (syoo'tyoor; "seam" L). The interwedged line of the suture is such that the bones cannot be separated short of breaking them. The cranium in the adult is, therefore, to all intents, a single bone.

As for the face, though it possesses a smaller surface area than the cranium, it possesses more bones—almost twice as many. These include 7 pairs of bones, plus one that is unpaired, making 15 in all. To begin with, there are the two *nasal bones* ("nose" L) which form the bridge of the nose and meet at the midplane. Behind the nasal bones are the *vomer bones* (voh-mer; "plowshare" L), named for their shape, of course, but making a comparison that is lost on today's urbanized population. The vomers make up the bony portion of the tissues dividing the interior of the nose into two nostrils.

Such a partition is a *septum* ("partition" L). The lower portion of the septum is not bony, but is cartilaginous, as you can tell from the manner in which you can bend and twist it. The fact that the human skull contains nothing to indicate the portion of the nose we are most aware of, gives the skeleton a horridly snubbed appearance that lends it its ugliness—along with the empty eye sockets and the ghastly grin.

The rear of the nostrils is bordered by the *inferior nasal conchae* (kon′kee; "shell" L, because it has a spiral shape something like that of a snail's shell). There are also middle and superior nasal conchae; they are not separate bones, but, rather, are processes of the ethmoid ("inferior" refers, remember, to something which is below; "superior" to something above).

Behind the nasal bones, and making up part of the orbit, are the *lacrimal bones* (lak′ri-mul; "tear" L). These are so called because they are located in the neighborhood of the tear ducts.

Thus there are no less than eight bones making up the nose and its environs; the two nasals, the two vomers, the two lacrimals, and the two inferior nasal conchae.

Most of the front of the face from the eye to the upper jaw is stiffened by the *maxillary bones* (mak′si-ler′ee; "jaw" L). These bones meet at the midplane and make up the entire upper jaw, which is therefore referred to as the *maxilla* (mak′si-luh). It also makes up part of the upper border of the mouth, runs under the cheeks, and reaches up to the eye, forming part of the orbit. Behind the maxillaries in the roof of the mouth are the much smaller *palatine bones,* also meeting at midplane, so called because they make up the palate—that is, the rear of the roof of the mouth. The *zygomatic bones* (zy′go-mat′ik; "yoke" L, another reference to a shape that is no longer familiar to many people) make up the sides of the face to the front of the sphenoid and temporal bones of the cranium. They form the bony overhang above the upper jaw and are therefore popularly called the "cheekbones." The zygomatics also reach to the border of the eye and make up part of the orbit. All told each orbit consists of portions of no less than seven bones of the face and cranium.

All the bones of the face I have mentioned so far are immovably joined to each other and the cranium, so that the skull, down to the line of the upper jaw at least, is a single piece. But there is one more bone in the head and it is the one movable bone. Naturally, I refer to the lower jaw.

As I have said in Chapter 1, the vertebrate jaw was originally formed in the placoderm out of the first gill arch. Originally the jaw, so formed, was separate from the remaining skeleton of the head. It still is in the sharks. However, in the bony fish, the upper jaw fused with the cranium, and this is the situation with all their tetrapod descendants. The lower jaw remains hinged in the rear to the upper jaw and must remain movable, of course, if biting and chewing is to be kept in the realm of possibility.

Here, too, the evolutionary trend has been in the direction of a decrease in the number of bones and a consequent strengthening of structure. The numerous bones in the lower jaw of the reptiles were reduced to two, one on each side, in mammals and these two are fused into a single piece by the second year of life in the human being. This lone bone of the lower jaw is the *mandible* ("chew" L).

The human being (and mammals, generally) have not lost all trace of those other bones of the reptilian lower jaw. As the mandible expanded in size and virtually shoved the other bones backward, several ended in the middle ear as the *ossicles* ("little bones" L). These are six in number, one set of three being found in each middle ear. They are named for their shapes: *stapes* (stay'peez; "stirrup" L), *malleus* (mal'ee-us; "hammer" L), and *incus* (ing'kus; "anvil" L). Of these, the stapes is thought to be a remnant of the second gill arch rather than the first, of which almost all the jaw is formed.

The ossicles are not generally counted among the bones of the skull, and neither is another bone set at the base of the tongue, the *hyoid bone* ("U-shaped" G). This is also a remnant of the second gill arch. Although it is sometimes called the "tongue bone," it is not in the tongue itself; it lies between the mandible and the voice box, and is unusual in that it is jointed to no other bone and lies in isolation. In fish, the bone was an important link between the jaw and the rest of the cranium, but it has lost that function in us.

THE TEETH

Set in both upper and lower jaws in man are the various teeth. These are *not* bones. They are hard, to be sure (harder than bone, even), and are constructed chiefly of

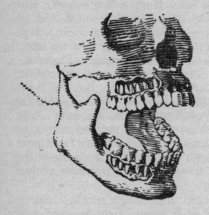

calcium phosphate. Nevertheless, the structure of teeth differs considerably from that of bone.

Teeth originated among the primitive sharklike fish and probably represented modified fish scales to begin with. At first, they were undifferentiated, all being of the same simple conical shape. There were many of them and they were replaced when worn down. The evolutionary tendency, however, was to reduce the number of teeth and the number of times they could be replaced. Furthermore, there was a change in the direction of greater efficiency through the specialization of groups of teeth for different functions.

Among the higher mammals, the number of teeth has been reduced to a maximum of 44; this maximum is found in dogs and pigs, for example, although in many other animals (and in man) the number is well below the maximum. Mammalian teeth fall into four different groups, all of which are represented in our own mouths.

In front of the mouth are the *incisors* ("to cut into" L). These are wedge-shaped teeth with the cutting edge compressed into a narrow line, so that when the incisors above and below meet they have a scissors action. To accomplish this most efficiently, when the jaw closes the lower incisors should come up just behind and in contact with the rear surface of the upper incisors. If the upper incisors are set too far forward, or the lower ones too far backward so that there is a gap between them when the jaw closes, most of their efficiency of action is destroyed. This is one of the more common varieties of *malocclusion* (mal'o-kloo'zhun; "bad closing" L).

Behind the incisors are the *canines* ("dog" L), conical tearing teeth that are the least specialized and most nearly resemble the ancestral shape. They are prominent in dogs, as one can deduce from the name. The canines of the upper jaw are often called "eye teeth," a name that arose in the mistaken impression that their roots were more or less intimately connected with the eye.

Next we have the *bicuspids* ("two-pointed" L), which have the appearance of double canines, since they seem to consist of two cones fused together. The double point that results gives them their name.

Finally, there are the *molars* ("millstone" L), the working edges of which have four or five blunt cusps, or points, that fuse into an uneven surface. These exert a grinding millstone-like action on food, since the lower jaw works from side to side. These teeth are often called "grinders" in common speech as a consequence. The bicuspids, coming before the molars, are very commonly called *premolars*.

Among the full 44 teeth possessed by higher mammals, 12 are incisors, 4 are canines, 16 premolars and 12 molars. The number of teeth are symmetrical on either side of the midplane of the face, so that in listing the teeth of any particular species it is only necessary to number them on one side of the mouth, the other side being presumed identical (barring loss of teeth through disease or accident, of course). However, the upper and lower jaws are not necessarily identical, so both are described. The numbers are placed in the order in which they occur in the mouth: incisor, canine, premolar, and molar. The dental formula for the horse or pig can be presented thus:

$$\frac{3 \cdot 1 \cdot 4 \cdot 3}{3 \cdot 1 \cdot 4 \cdot 3}$$

However, not only is the number of teeth reduced in many mammals but a varying emphasis is placed on one group or another. In rodents, the incisors are enlarged and are by far the most prominent teeth in the mouth. They are permanently growing, moreover, being replaced by new growth as fast as they are worn down by the gnawing activity of the animal. In carnivorous animals, the canines are much enlarged, forming, as an example, the "fangs" of a tiger. Again, in grazing animals such as cattle and horses, which must continuously grind coarse grains and grass, the molar teeth are much developed and have an exceedingly intricate grinding surface.

Specialization sometimes emphasizes teeth to almost grotesque extent, as in the tusks of elephants (which are overgrown upper incisors) or those of walruses (which are overgrown upper canines).

On the other hand, teeth that are not useful to an animal's way of life may be suppressed altogether. Cattle lack incisors and canines in the upper jaw, and the sperm whale entirely lacks teeth in the upper jaw. The narwhal possesses only two teeth all told, one of which (in the male) grows forward to form a spiral tusk up to eight feet long. Anteaters possess no teeth at all, and this is true of the entire class of Aves (birds).

The teeth of the human being are far less specialized than the teeth of most mammals, in the direction of either over- or underdevelopment. (As a matter of fact, the structure of the human being is surprisingly primitive and unspecialized in comparison with that of most other animals.

This is, perhaps, a clue to our success, for we have not committed ourselves too far in any one direction.)

To be sure, human teeth are small, considering our size and weight, but that is part and parcel of the general shrinkage of man's face over a million years of evolution. In most animals the face is drawn forward into a "muzzle" so that the jaws may seize while allowing the eyes still to see. The jaws are therefore large enough to accommodate large teeth. The development in man (and in related creatures) of hands that can seize food and bring it up to the mouth made the muzzle unnecessary. Smaller jaws could accommodate only smaller teeth.

This is also one reason why the teeth of the human adult fall short of the full mammalian number by twelve. And yet, despite this loss, we have retained at least some of each variety of tooth, have emphasized none of them unduly, and have kept the number the same in upper and lower jaw. The dental formula for man is:

$$\frac{2 \cdot 1 \cdot 2 \cdot 3}{2 \cdot 1 \cdot 2 \cdot 3}$$

Mammals generally possess two sets of teeth in their lifetimes. The reason for this is clear. The jaw of the young mammal is simply too small to accommodate the size and number of teeth the adult will require. Nor can we expect teeth to erupt in small size and grow with the child, for once teeth have erupted, they lack the capacity for further growth.

The human child, for instance, makes out at first with 20 small teeth. These are variously called *deciduous teeth* ("to fall off" L, since that is what they will inevitably do), "temporary teeth" (which gives the same idea), "milk teeth" (because the infant gets the first of them when it is still very largely on a milk diet), and "baby teeth" (which is obvious).

At birth, these teeth are already forming within the gums, but the first of these, the two lower middle incisors, do not actually cut through the gums until the second half year of life. (The process is fairly painful and during the period of "teething" the child becomes fretful and a great trial to its parents.) When the child is two or two and one half, the process may be complete, at which point the dental formula is:

$$\frac{2 \cdot 1 \cdot 2 \cdot 0}{2 \cdot 1 \cdot 2 \cdot 0}$$

As you can see, the incisors, canines, and premolars are of the same number as in the adult. It is the 12 molars that are missing. Though the 8 rearmost baby teeth are called molars, they are replaced by adult premolars. The true adult molars come in fresh and without predecessors.

The first of the "permanent teeth" or "adult teeth" to come in are the first molars, at about the age of six. They enter behind the baby teeth, in a jaw which, by then, has grown large enough to have room for them. After that, the baby teeth begin falling out, starting at the front and working backward. There is usually a lapse of time between the going of the first teeth and coming of the second, which gives rise to the characteristically toothless smile of the six- or seven-year-old.

By the time the child is twelve, the first teeth are entirely replaced, and it is only in the teens that the second and third molars come in, the jaw by then having grown almost large enough to accommodate them. I say "almost" because, as a matter of fact, the human jaw does not generally grow large enough to accommodate the third and final set of molars comfortably. Eruption of these is commonly delayed until the age of 20 or even later, sometimes, as though to give the jaw a last chance, so to speak. For this reason, the third molars are popularly known as "wisdom teeth," since they make their appearance only at an age when the owner may be considered (optimistically) to have attained to years of wisdom.

On occasion, one or more—even all four—of the wisdom teeth may not erupt at all. This is no great loss, perhaps; on a modern human diet eight molars will do. Then, too, in jaws where the wisdom teeth do erupt, they are often uncomfortably crowded and may even be so firmly wedged ("impacted") between the jawbone and the second molar that their removal, when made necessary by the decay to which they are all too prone, becomes a matter of almost major surgery.

There is good reason to think that the wisdom teeth are on the way out and that in a relatively short time (evolutionarily speaking) man's teeth will be reduced in number to 28.

3

OUR LIMBS AND JOINTS

THE ARMS

The skull, vertebral column, ribs, and sternum, all taken together, represent the *axial skeleton,* forming as it does the axis of the body. Evolutionarily speaking, this was the original skeleton. The bones of the limbs and the structures related to them form the *appendicular skeleton* ("to hang from" L), since the limbs do indeed, in a manner of speaking, hang from the body proper. They are "appendages," in other words. Originally the appendicular skeleton was small in comparison with the axial skeleton, for, when first appearing in the late placoderms and early sharks, they were needed only for the bracing of stubby fins.

Among tetrapods the limbs had to become larger and stronger in order to support the body against gravity, and this tendency continued among mammals. Longer limbs raised the body, together with the head and sense organs, higher off the ground and made the field of vision larger and the opportunities for long-distance hearing and smelling greater. Furthermore, the longer the leg the faster the movement of the extremity for a given angular motion, so that a long-legged animal can run faster than a short-legged one, a matter of value both in pursuit and escape. (This holds true for nonflying birds as well—consider the long-legged ostrich.)

Man shares this tendency with the mammals generally so that our legs are longer than the torso itself and more bones are to be found in the appendicular skeleton than in the axial skeleton. It is to the bones of the legs moreover that the variations in human height are largely due. The human spine averages 28 inches in length in the male and 24 inches in the female, with surprisingly little variation from individual to individual. It is the length, or lack of it, in the bones of the legs that is responsible for most of the variation in height. You can see this for yourself if you observe a group of seated men who all stand up and suddenly become much more heterogeneous in height.

Among the various tetrapods, the limbs have undergone a variety of modifications suiting the way of life of the particular creature. In the case of mammals that have returned to aquatic life, the limbs have reverted to an almost fishlike stubbiness and have become paddles. (In whales and sea cows, the hind limbs have completely disappeared, at least as far as any external evidence is concerned.)

In the case of birds and bats, the forelimbs have been modified into wings, and in instances where birds have become flightless those wings shrink and (in at least one case, that of the New Zealand kiwi) have just about disappeared. Animals like the kangaroo that hop or jump as the favored means of locomotion developed oversized hind limbs and those that swing through the branches of trees, like the gibbon, have developed oversized forelimbs.

All, however, retain much the same basic bony plan. It is the basic similarity of the bones of the human arm, whale's flipper, bat's wing, and bear's leg which is one of the most striking manifestations of the close relationship among vertebrates. It is almost impossible to look at such a pattern of similarity without imagining a long gone ancestor that supplied the basic theme on which modern species only ring variations.

And in the case of limbs, as well as teeth, the human being is comparatively unspecialized; except for the lengthening of some of the bones, the arms especially remain

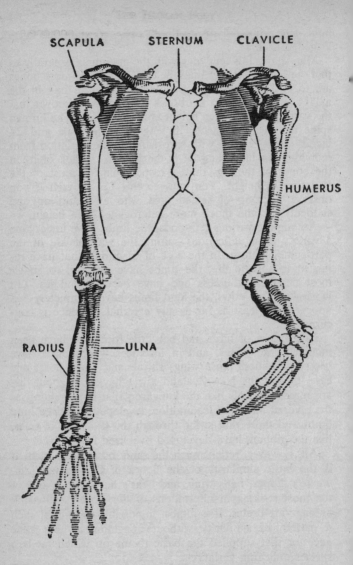

remarkably similar to what the original tetrapod limb must have been like.

The bones of the arm are connected to the axial skeleton by way of two pairs of bones in the upper torso, one in front and one in back. Those in back are the "shoulder blades," broad flat bones which, if you twist your arm behind your back, stand out under the skin like burgeoning wings. The formal name is the *scapula* (skap'yoo-luh; "to dig" G), because of the resemblance of the bone to the business end of a spade. The scapulae lie dorsal to the ribs but do not actually make contact with them, since there is a binding layer of muscle between.

The pair of bones in front, ventral to the rib cage and just above the first rib, are the "collarbones." You can feel them at the base of your neck just about where your collar is. They are long narrow bones, gently curved like an italic lower-case *f*. To some imaginations it is shaped like an old-fashioned key, a point commemorated by its formal name *clavicle* ("little key" L). The end of the clavicle near the midline of the body adjoins the upper end of the sternum; at the other end it adjoins the scapula. (The clavicle is important to the mythology of childhood, for in birds this pair of bones join firmly to form the familiar V-shaped "wishbone.")

If you were to look down at these bones from above, they would seem to form a double crescent almost encircling the upper part of the body. It is not a complete circle, since there is a short gap of about an inch between the two clavicles and a larger gap between the two scapulae. However, if you count the stretch of sternum and vertebrae between, you can consider the bones as a girdle, the *pectoral girdle* (pek'to-rul; "breast" L), in reality.

Adjoining the pectoral girdle are the bones of the arm itself. The arms are divided into three segments, an upper arm, a lower arm, and a hand. The leg and, actually, the tetrapod limb in general are similarly divided into three portions.

In describing the structure of the limbs, the adjectives *proximal* ("nearest" L) and *distal* (a word coined from "distant") become useful. That section of limb or of any extended structure which is nearer the trunk or the midplane of the body is the proximal part. The opposite end would be the distal part. Thus, the upper arm is the proximal portion and the hand is the distal portion. The lower arm is the intermediate portion.

In all tetrapods, the proximal portion of the limb contains a single bone, while the intermediate part contains two. This is certainly true of the human arm. As the limbs lengthened, these bones lengthened and became the "long bones" of the body.

The upper arm contains one long bone, the *humerus* (hyoo'-mur-us; "shoulder" G).* The lower arm contains the two long bones, the *radius* ("ray" L) and the *ulna* ("elbow" L). A ray obviously is something that radiates outward from a center and the word was originally applied to the spokes of a wheel. The radius of the lower arm apparently seemed straight enough to be such a spoke; whence its name. The ulna is likewise appropriately named —the bony portion of the elbow is indeed the end of the ulna.

The distal end of the limb contains many bones; 27, as a matter of fact. This situation dates back to the origin of the limbs when their stubbiness was braced by a number of small, irregular bones. There was a value to this. A single bone would have made a flipper a stiff and inefficient oar. A single line of bones would have allowed it to bend as our vertebral column bends but only as a unit. A number of bones spread out in two planes—lengthwise and breadthwise—and able to slide past one another to a limited degree, introduce a two-dimensional flexibility and allow the delicacy of maneuver required for efficient steering. Three of the bones lengthened out to form the upper and lower limbs in the extended version required by tetrapod structure, but the distal end of the limbs retained the remaining bones.

When the first amphibians clambered onto the muddy tidal flats and took to living on land, they needed not only stronger support for their limbs but also a more splayed-out surface where limb met ground, to keep them from

---

* It is almost irresistible to wonder if the humerus is by any chance the "funny bone" which gives us that unpleasant tingling feeling when we jar our elbow. To be sure, the humerus ends near the point we jar, but the source of the sensation is not a bone at all, but a nerve, and the relationship between "humerus," "humorous," and "funnybone" is merely coincidence.

sinking into the mud—a sort of snowshoe effect. For this purpose, the small bones at the distal end of the limb spread out (probably with webs between), distributing body weight more widely and evenly upon the mud. Each limb developed a number of bone-braced digits and the primitive number was five on each limb. There seems no particular reason why the number should have been five, but that is what it was and no normal living tetrapod has more than five digits on any limb.

With the passage of time, the evolutionary tendency often lay in the direction of a reduction in the number of digits. On hard ground there is less necessity for a splayed-out limb. It is more important, instead, to develop a fleshy pad or horny sheath to absorb shocks and make possible the pounding that would accompany a fast run. Where the pad develops, the digits decrease in size and become mere claw-carrying devices; as in the cat. Where horny sheaths ("hoofs") develop, the number of digits tends to decrease so that the individual hoof can become larger and stronger. The rhinoceros has only three toes left; cattle, deer, and ruminants generally have only two; the horse and related creatures carry the process to its logical conclusion and have but one horn-sheathed toe on each leg.

It is the sign of the primitive character of the human arm that it retains five long, maneuverable digits. Of course, it is not for support on marshy ground that we need them. Rather, we have turned the hand into a superlative manipulative organ, incomparably the best thing of the sort in all the realm of life—with four limber fingers and an opposing thumb so that the whole can be used as a delicate pincer or firm grasper, a twister, bender, puller, pusher, and manipulator of piano and typewriter keys.

The bony condition of the original flipper persists in the wrist, where eight irregular bones in, roughly, two lines of four each (all in close contact) make for flexibility. The wrist can bend backward and forward easily and, to a more limited extent, to the left and right.

The eight bones as a group are the *carpal bones* ("wrist" L), and individually they are named after whatever it was that the particular bone seemed to resemble in the colorful fancy of early anatomists. They are the *navicular* (na-vik'yoo-ler; "boat-shaped" L), *lunate* ("crescent-shaped"

L), *triquetrum* (try-kwee'-trum; "three-cornered" L), *pisiform* (py'si-form; "pea-shaped" L), *greater multangular* ("many-angled" L), *lesser multangular, capitate* ("head-shaped" L, that is, rounded), and *hamate* ("hooked" L).

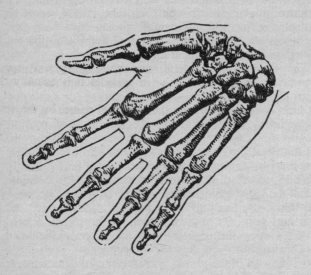

The hand itself contains 19 bones, arranged in five rows. Four of the rows contain four bones, the fifth contains three. The five bones that adjoin the carpals are the *metacarpal bones* (met'uh-kahr'pul; "after the wrist" L). They are encased in soft tissue, and form the palm of the hand, but it is easy enough to feel the five separate bones under the skin of the back of the hand. The metacarpals are numbered one through five, beginning at the thumb. The second, third, fourth, and fifth metacarpals are virtually parallel and immovable; but the first is set at an angle and has limited mobility.

Adjoining the metacarpals are the "fingerbones," or *phalanges* (fuh-lan'jees), of which the singular is *phalanx* (fay'lanks). The phalanx in Greek history was a close array of soldiers side by side and back to front. The bones of the fingers, set in similar close array, were reminiscent of that, and this accounts for the name. Each finger, except the thumb, has three phalanges of decreasing size as one proceeds distally. The thumb has but two. Some anatomists feel that the thumb can be counted as having three phalanges, too, if the first metacarpal is considered a phalanx. If that were so, then the first phalanx of the thumb would connect with the carpals directly and there would be only four metacarpals instead of five.

(The absence of flesh in a skeleton allows the meta-carpals to add their separateness to that of the phalanges, and this gives the skeletal hand its long-fingered aspect. What seems to be the palm in a skeleton is actually the wrist.)

Each finger has a formal anatomical name, by the way. The thumb is *pollex* ("strong" L) because it is stronger than the other fingers, as you recognize whenever you push a tack into a board with the thumb, rather than with any other finger; hence a "thumbtack." The other four fingers are, in order, *index* ("pointer" L), *medius* ("middle" L), *annularis* (an'yoo-lar'is; "ring" L), and *minimus* ("least" L).

THE LEGS

The legs, which in bipedal man bear the brunt of support and locomotion, are longer, stronger, and more specialized than the arms. Yet the similarities, indicating the common plan of all four limbs, are unmistakable. To begin with, just as there is a pectoral girdle so there is another girdle (far heavier and stronger), to which the structure of the leg is attached. This lower girdle consists of three pairs of bones: the *ilium* (il'ee-um; "groin" L), the *ischium* (is'kee-um; "hip" G), and the *pubis* (pyoo'bis; "adult" L). One of each of these is on either side of the midplane, and all together form the bony structure of the hips.

The ilium and ischium are flat, irregular bones—the ilium above (superior) and the ischium below (inferior). You can feel the crest of the ilium on either side of the body just below the waistline. It is upon the ischium, on the other hand, and on the muscles attached to it, that you sit.

In front, smaller than either ilium or the ischium, is the pubis. It joins the ischium in such a way as to form a pair of large holes at the bottom of the girdle which are very prominent when the skeleton alone is viewed. These are the *obturator foramina* (ob'tyoo-ray'ter; "stopped-up holes," because in actual life they are almost entirely covered by a membrane). The derivation of "pubis" from "adult" arises in the following manner. A common sign of adulthood, or at least of sexual maturity, is the appearance of hair in the genital region. Such hair is therefore the "pubic hair" ("adult hair") and the bone which is to be found just under that region came to share the name.

In the child these three bones are separate, but by the middle twenties, they are firmly fused into a single bone, which in common parlance is the "hipbone." A more formal name is *os coxae* (os kok'see; "hipbone" L). Before this rather obvious name,—for Latin-speakers—was accepted, the name most often used, and still much used, was *os innominata* (i-nom'i-nay'tuh; "nameless bone" L) because, though the three parts had names, all three together had not. ("Innominata" is also applied to other anatomical parts without names, so that the very namelessness becomes a name.)

The two os coxae meet in front, where the pubes join by way of a cartilaginous link something like those between vertebrae. This is the *pubic symphysis* (sim'fi-sis; "a growing together" G). Dorsally the two ilia do not meet. Rather, they join the sacrum, one on either side. (The fusion of the five sacral vertebrae thus makes for a stronger hip structure.) So neat and firm is the connection between sacrum and ilium that it is common to speak of them as though they were a single bone, the "sacroiliac." Because of man's troubles in the small of the back resulting from the imperfections of bipedal structure, the word has come to have most unpleasant connotations.

The hipbone and sacrum, taken together, form a complete bony girdle, stronger in proportion to man's size than are the similar structures in other mammals. (This is not surprising, in view of man's bipedal position.) Furthermore, in no other mammal do the hipbones form such a rounded and basin-like structure. Again, this is the consequence of bipedality. In a four-legged creature,

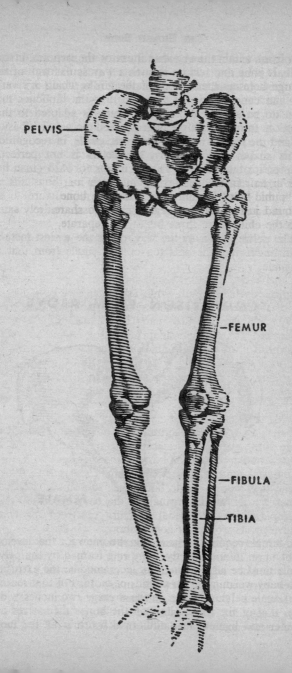

the organs within the abdominal cavity are suspended from the backbone and rest on the ventral muscular wall of the abdomen. In man, the abdominal wall is vertical (or should be) and cannot serve as support. It is the hipbones that must do so and their basin-like shape is adapted to that function. In fact, the hip region is called the *pelvis* ("basin" L) and the ring of bone is the *pelvic girdle,* in recognition of its shape. Unfortunately, the pelvis is not perfectly adapted for the purpose. The basin tips forward (man has only been at bipedality some hundreds of thousands of years and it takes more time to adjust structure to so radical an innovation) and the support isn't entirely satisfactory.

The pelvic girdle, by the way, offers the easiest method of distinguishing the skeleton of the female from that of the male.

COMPARISON FROM ABOVE

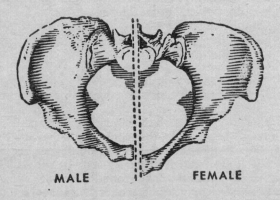

MALE   FEMALE

The female requires room in the abdomen for the development of an infant, and the bony ring formed by the pelvic girdle must be large enough to accommodate the extrusion of a baby weighing seven pounds or more. For that reason, the female pelvic girdle is on the average two inches wider than that of the male, although the bones themselves are thinner and lighter. This additional width is all the more obvious because of the generally smaller size of the rest of the skeleton of the female as compared with that of the male.

Further, the angle made by the meeting of the two pubic bones at the symphysis is much wider in the female, where it is about 90 degrees, than in the male, where it is only 70 degrees. The net result of all this is that the living female possesses prominent hips which may be a cause for concern to her at those times when slenderness is in fashion. They are, however, essential to the smooth functioning of her role as a mother, and, thanks to the inscrutable wisdom of Nature, prove attractive to the male of the species, whatever the fashion.

Adjoining the pelvic girdle is the *thigh* (AS), which is the proximal portion of the leg. This, like the proximal portion of the arm, contains one long bone. This "thighbone" or *femur* (fee'mur; "thigh" L) is the longest bone in the body, making up about ⅖ of the height of an individual. At its proximal end it has a very characteristic rounded head, which is set to one side and fits into a rounded depression, or socket, in the hipbone. This socket is the *acetabulum* (as'i-tab'yoo-lum), so called because of its resemblance to a round cup the Romans used for holding vinegar (*acetum*).

The intermediate section of the leg, again like that of the arm, contains two bones. However, whereas the two bones in the arm are nearly equal in size, those in the leg are quite unequal. The larger of the two is the *tibia* (tib'ee-uh; "flute" L; which in length and shape it seems to resemble). The tibia, commonly called the "shinbone," is the second longest bone of the body. It runs just under the skin in the forepart of the leg and you can easily feel it. Its distal end, at the ankle, forms a swelling, which you can feel as the bony protuberance at the inside of the ankle.

The smaller bone of the intermediate section is as long as the tibia but is much thinner, so that it is commonly called the "splintbone," since it resembles a splinter broken off the main bone. It is, in fact, the thinnest bone, in proportion to its length, in the body. Its formal name, *fibula* (fib'yoo-luh; "pin" L), also indicates this: the implication is that its relation to the tibia resembles that of a pin to a brooch. For most of the length of its course, the fibula is well hidden under muscle and cannot be felt by probing fingers. At its distal end, however, it forms the bony prominence on the outside of the ankle.

The knee, which is the joint between the proximal and intermediate sections of the leg, differs from its analog, the elbow, in that it possesses a separate bone. This is the small, flat, triangular "knee bone," or *patella* (puh-tel'uh; "small pan" L). It protects an important joint which in the ordinary course of walking, and particularly of running, is constantly being pushed out ahead of the body. Like the hyoid bone, the patella is not directly connected to any other bone, although it is held in place by muscle attachments. If you relax your leg muscles so that the patella is not held firmly in place by them, you will find that you can move it about quite a bit in any direction.

The distal section of the leg contains the ankle and foot, which are analogous to the wrist and hand. The ankle, like the wrist, contains a series of irregular bones, though it possesses only seven as compared with the eight of the wrist. The ankle bones make up the *tarsus* ("wicker-basket" G, the name apparently being suggested by the fact that the separate bones in such close association resembled the interwoven wicker strands of such a basket). One of the bones of the tarsus, the *calcaneus* (kal-kay'nee-us; "heel" L), does indeed, as the name implies, extend backward to form the heel. It is the largest bone of the tarsus.

This backward extension of the calcaneus would seem an attempt to make man's bipedal support more stable. Any object resting upon two narrow supports is at best in unstable equilibrium, and any jar will topple it. By extending a heel backward, the support is propped up and even made somewhat quadrupedal. A man stands not on two feet, but on two soles and two heels. It is not much of an advance toward stability—it is still easy for a man to fall (and, in fact, a baby learns to walk only through an inevitable succession of falls)—but it answers the purpose. A careful adult can go sometimes for years without falling and yet without seriously limiting his locomotion, either.

The remaining six bones of the tarsus are the *talus* (tay'lus; "ankle" L), *cuboid* ("cube-shaped" L), *navicular* ("boat-shaped" L), and the first, second, and third *cunei-*

form (kyoo-nee′i-form; "wedge-shaped" L). The calcaneus and talus bones are roughly cube-shaped and so, of course, is the cuboid. The soldiers of ancient Rome used such bones (obtained from horses, usually) to hack out rough-and-ready dice. For this reason, the talus in particular is sometimes called the *astragalus* (as-trag′uh-lus; "die" L).

Just as the hand is made up of metacarpals and phalanges, the foot proper is made up of the five parallel bones of the *metatarsus* plus five sets of phalanges. As in the hand, the first digit—the big toe, or *hallux* (hal′lux; "big toe" L)—has two phalanges, the other digits three.

The foot is one of the more specialized portions of the human skeleton. It has gained a heel and has lost those characteristics of the hand that the feet of our early an-

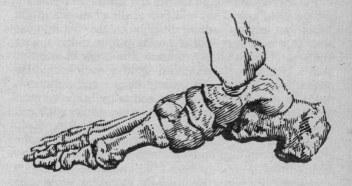

cestors must have had. Living apes and monkeys still possess hind feet with much of the characteristics of hands. Those animals have opposable big toes, well separated from the other toes, so that the foot can be used to grasp. Indeed, the apes and monkeys are sometimes lumped together as the *Quadrumana* (kwod-roo′muh-nuh; "four-handed" L).

In man, as another token of specialization, the foot possesses a big toe that is parallel to the other digits and is not opposable. The phalanges have shrunk almost to nothing in the other toes so that the foot is almost one piece. In man, each pair of limbs has its special job; the hands are the graspers and the feet are the support; whereas in the other primates, both limbs compromise their function.

The use of two limbs alone as a support affects us in another way. Most mammals have cut into the support function of the legs by raising themselves to their toes, thus adding the length of their metatarsals to their height. Such creatures are *digitigrade* (di'ji-ti-grayd'; "toe-walking" L), and the best-known examples are the cats and dogs. The additional height, with its advantages of better placement of the sense organs and greater speed, is counterbalanced by the smaller area of the feet touching the ground and the consequent increase in the stresses upon the feet. Apparently the advantages outweigh the disadvantages in their case.

The hoofed animals go further and lift all the phalanges but the distal one. They add two of the phalanges to the height and end up walking literally on the tips of their toes. This is all very well if you have four widely-spaced supports. It is possible to narrow the areas of contact. Man, with but two supports, can afford no such luxury. He must splay them out and keep phalanges and metatarsus firmly on the ground. If he must add to his height he must do so by lengthening only the long bones of the thigh and shin. He is *plantigrade* (plan'ti-grayd; "sole-walking").

There are other plantigrade mammals, too—notably the bear, which can take up an erect posture better than most tetrapods. Man goes the bear (and also other plantigrade mammals) one better by using the calcaneus, so that he actually walks on his ankles, in part.

The sole of man's foot is not straight for the same reason that the vertebral column of a tetrapod is not. We need an arch for strength, and we have it in the sole. Thus weight is transmitted to the heel and to the ball of the foot, and there is also an elasticity to absorb the shock of walking, when weight is constantly being shifted from foot to foot. (Here is another important specialization of the human foot. Apes do not have arches.) But here again, adjustment to bipedal postures is not perfect. The structures forming the arches can give under weight and flatten out. The "flat feet" that result lower the efficiency of walking and can make it downright painful after a while as the successive jars, relatively unabsorbed, are transmitted to the spinal column and the skull.

CELLS

So far I have been describing the external appearance of the bones and the position of each in the body. This gives a rather static impression of the bones as an inert framework and nothing more. To be sure, a hard mineral makes up 45 per cent of the weight of bone and this part is dead; but in life, bone is more than its mineral matter and is anything but inert. Within its mineral makeup, and within the structure of cartilage, too, are living cells.

The *cell* is the unit of living tissue. It received its name in 1665, when the English scientist Robert Hooke, one of the first microscopists, observed that a thin slice of cork showed a spongy structure and contained tiny, oblong, regularly-spaced holes. The name "cell," meaning a small room, seemed ideal for those holes. The equivalent Greek word is *kytos* and this is often used in such compound words as *cytology* (sy-tol'uh-jee),* meaning the study of cells.

However, the holes observed by Hooke were but dead remnants of a woody skeleton. In living tissue there is the same spongy structure but the cells are not empty. They are filled with a gelatinous material which in the early 19th century received the name of *protoplasm* ("first form" G).

The cells are complicated indeed in structure, but for purposes of this book only the barest description will be necessary. In the first place, cells are small. The largest cell in the human body is the egg cell produced by the female, about the size of a pinhead and just visible to the naked eye. Other cells in the body fall far below this in size and can be seen only by microscope.

Each cell has a boundary, a thin and delicate *cell membrane*. The membrane marks off the interior of the cell from the outer environment; and the chemical and physical structure of the regions on either side of the membrane are completely different. There is a natural tendency for struc-

* The Greek *k* becomes the Latin *c*, which is always hard but in English often becomes soft. Hence the switch from the Greek *k* sound to the English *s* sound.

ture and composition to equalize across the membrane; but it is the essential function of life to maintain the difference despite the equalizing tendency. The membrane is only about 10 millimicrons thick (a millimicron is one billionth of a meter, or 1/25,000,000 inch) and consists of only three or four layers of complex molecules. Nevertheless, it somehow serves as a selective and oneway passage for certain substances from environment to interior and for others from interior to environment. The mechanism by which this is done is by no means understood as yet.

Within the membrane the cell is divided into two major portions. A small central region called the *nucleus* ("little nut" L, because it resembles a small nut within a larger shell) is surrounded by its own *nuclear membrane*. The nucleus controls cell division and contains the mechanism that ultimately dictates the nature of the cell's chemical machinery.* Between the nucleus and the cell membrane is the *cytoplasm,* which carries on the everyday work of the cell.

The cell is complicated enough to serve not only as a component of living tissue but as an independent organism. There are numerous species of one-celled organisms. Nevertheless, all plants and animals that we see about us with our unaided eye are made up of a number of cells. The human body contains more than 50,000,000,000,000 (50 trillion) cells.

In a multicellular organism the cells fall into specialized groups, each performing a particular function with particular efficiency, to the exclusion, sometimes, of adequate performance in other functions equally vital to life. This means that an individual cell from a multicellular organism cannot maintain its life independently but only as part of a complex group, where other cells make up its deficiencies and where a smoothly working organization unites and controls all the specialties. (The analogy of a modern society containing numerous highly specialized human beings who would quickly starve if marooned separately

---

* This is a tremendous subject with which there is no room to deal in this book. If you are curious, you will find a rather detailed discussion of such matters in my book *The Wellsprings of Life* (1960).

on uninhabited islands but who can get along perfectly well within the social structure, is irresistible.)

A particular tissue is made up of a mass of cells of similar specialization. Those cells which specialize in forming the various substances that in one way or another hold the body structure together make up *connective tissue*. The specialized function of the cells of the connective tissue is to manufacture about themselves those molecules that make up bone, cartilage, and other portions of the connecting framework of the body.

Many of the molecules so manufactured are *organic* in nature; that is, made up chiefly of the elements carbon, hydrogen, oxygen, and nitrogen, which constitute the major portion of all living tissue. Such molecules are in contradistinction to those which lack carbon (the key element of life) and which are similar in properties to the substances making up the nonliving air, sea, and rocks about us. These latter compounds are, very naturally, called *inorganic* ("not organic"). Despite the latter name, the body can and does make use of inorganic substances. Water is inorganic and so is the calcium phosphate that makes up the major portion of the bones.

The organic substances of the connective tissue fall into two classes: *protein* and *mucopolysaccharide* (myoo′koh-pol-ee-sak′uh-ride). The proteins are particularly complex molecules, built up of long chains of smaller molecules called *amino acids* (a-mee′noh). A single protein molecule will contain thousands of atoms, even millions sometimes, arranged in helical coils, like miniature spiral staircases. The importance of protein to life is indicated by the fact that "protein" comes from a Greek term meaning "of first importance." In connective tissue the protein molecules associate themselves into bundles of coils—tiny fibers, that is—which mat together to form a sturdy fibrous structure that can be reasonably elastic, if the coils are properly arranged. The cells that produce this fibrous connective tissue are called *fibroblasts* ("fibrobud" G). The two chief proteins existing in connective tissue are *collagen* (kol′uh-jen; "glue-producer" G, because it will produce a glue if boiling is prolonged) and *elastin* (ee-las′tin, so called because of its elasticity).

The mucopolysaccharides are also large molecules, but

are built up of a series of units derived from the simple sugars. The "polysaccharide" portion of the name is from Greek words meaning "many sugars." A solution of mucopolysaccharide is gummy, viscous, and sticky, and the prefix comes from the similarly viscous material

FIBROBLAST

("mucus") secreted by many parts of the body. In fact, mucus possesses its properties because it is a solution of mucopolysaccharide.

One particular mucopolysaccharide is *hyaluronic acid* (hy'al-yoo-ron'ik), which occurs almost universally between cells and helps hold them together. It is sometimes called "ground substance" or "intercellular cement" for that reason. Another molecule of this type, containing some sulfur atoms in the molecule in addition to the more usual types, is *chondroitin sulfate* (kon-droh'i-tin). Cartilage is rich in mucopolysaccharides, and the Greek word for cartilage (*chondros*) is the source for the name chondroitin sulfate.

Cartilage is formed by relatively large oval cells called *chondrocytes* (kon'droh-sites; "cartilage cells" G) which form collagen and chondroitin sulfate chiefly, and deposit those substances outside the cell. The chondrocytes are thus separated by the cartilage they form, although they tend to remain in groups. Though the cartilage between the cells is nonliving, the cells themselves are alive.

The most common type of cartilage is *hyaline cartilage* (hy'uh-lin; "glassy" G) because of its clear and translucent appearance. (The occurrence of hyaluronic acid in such cartilage accounts for part of the name of that mucopolysaccharide.) It is as hyaline cartilage that most of the skeleton is first formed, and some of the skeleton remains in that form into old age, as, for instance, is the case with the costal cartilages connecting ribs and sternum.

There is also *elastic cartilage,* which is yellow in color. (Both elasticity and yellowness are due to its content of elastin.) Such cartilage occurs, for example, in the framework of the ear.

Finally there is *fibrocartilage,* in which molecules are bound together to form a tough fibrous substance rather than a soft elastic one. It is this fibrocartilage that forms the intervertebral discs and joins the two hipbones at the pubic symphysis.

CHONDROCYTE

OSTEOCYTE

BONE STRUCTURE

Despite the hard, dry appearance of bones in a skeleton on display, it is important to remember that in life about 25 per cent of bone weight is water and another 30 per cent is organic material. The organic material is almost entirely collagen, with some mucopolysaccharide also present.

Like cartilage, bone contains living cells whose function it is to manufacture the connective material. The difference is that the *osteocytes* (os′tee-oh-sites; "bone cells" G) also bring about the formation of mineral matter, which is then deposited in the organic framework, hardening it and giving it strength.

The mineral matter is chiefly a basic calcium phosphate, in which calcium ions are surrounded by phosphate ions and hydroxyl ions* in a pattern that is by no means unique to life. There are common minerals that show the same pattern. The closest approach is that of *fluoroapatite* (floo'ur-roh-ap'uh-tite) which differs only in containing fluoride ion in place of hydroxyl ion. For this reason, the mineral structure of the bone is sometimes spoken of as *hydroxyapatite* (high-droks'ee-ap'uh-tite). When bone has been buried for long periods of time in the soil, there is a slow tendency to replace the hydroxyl ion by fluoride ion, so the age of fossil bones can sometimes be judged by their fluoride ion content.

Bone also contains calcium carbonate in fair amount, together with small quantities of compounds of magnesium, sodium, and potassium. It is—in addition to being the rigid framework of the body—a complex mineral reservoir, its components being continuously available to the body.

Bone is penetrated by narrow *Haversian canals* (ha-ver'zhun, after the English physician Clopton Havers, who first described them in 1691). It is through these canals that the blood vessels and nerves pass. The osteocytes, ovoid cells with numerous jagged processes, are arranged in concentric layers about the canals. A Haversian canal and its concentric layers of cells and mineral is called an *osteon* (os'tee-on) and a number of osteons fuse together, looking like adjoining tree trunks under the microscope, to form bone.

The layers of mineral may be laid down quite densely to form *compact bone,* or mineral matter may be laid down in a series of separated bony fibers forming a spongy latticework. This is called *cancellous bone* (kan'sel-us; "lattice" L).

The long bones of the limbs show both forms of bone. The outermost surface is a layer of compact bone; within is cancellous bone. Such bones are lighter than they would

---

* In this book, I intend to involve myself as little as possible with chemistry. If you are acquainted with these ions, fine. If not, you can either find information on the subject in any elementary chemistry book or, if you prefer, you can ignore the matter and read on.

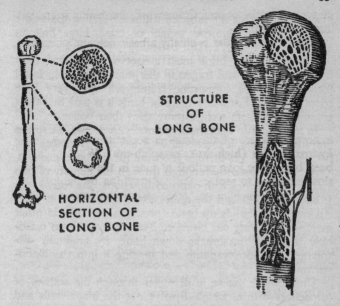

STRUCTURE
OF
LONG BONE

HORIZONTAL
SECTION OF
LONG BONE

be if they were compact throughout, and yet their strength is scarcely decreased. In the first place, a hollow cylinder is surprisingly strong. (A sheet of ordinary writing paper, rolled into a loose cylinder and bound so by a rubber band, will support a fairly heavy textbook.) In addition, the bony bars and plates within the cancellous region act as reinforcing struts formed along the lines of the tension and pressure produced by the normal movements of the body.

The hollowness of human bones is not unqualified. They are filled with a soft, fatty material called *marrow* (AS). The marrow is lighter than bone itself, and a marrow-filled hollow bone is lighter than a solid one (and uses up less inorganic material). Where lightness is particularly necessary, however, true hollowness will be found. The elephant, as an example, needs a huge skull upon which to base the muscles required to manipulate his massive trunk and to hold up a head weighted down by both the trunk and his majestic tusks. In order to supply the surface of

bone necessary for the muscles without negating the whole
purpose by a prohibitive weight of bone, large hollows
exist in the elephant's skull bones.

Similarly, flying birds must conserve weight and their
bones are hollow and fragile to the point where they fulfill
their function as supporting framework with very little
room to spare. For many types of birds it is true that their
coating of feathers weighs more than their bones.

In man, and in mammals, generally, however, there are
few true hollows in the bone. This has its advantages too,
for, as we shall see, there is good use for the bone marrow.

A bone grows or is repaired through the activity of two
kinds of osteocytes with opposed functions: the *osteoblasts*
("bone bud" G) and the *osteoclasts* ("bone breaker" G).
An osteoblast builds up bone (produces its buds, in other
words) by putting out layers of hydroxyapatite. An osteo-
clast is a cell that breaks down bone, by gradually dissolving the hydroxyapatite and feeding it into the bloodstream.

Thus a bone grows in diameter through the activity of
osteoclasts within, which dissolve out the inner walls and
make the interior hollow wider (but leave the strengthening
struts along the lines of pressure and tension). Meanwhile
osteoblasts are adding layers of hydroxyapatite on the outer
surface. In repairing a break in the bone, osteoblasts lay
down the minerals and osteoclasts polish off the rough
edges, so to speak, and remove the excess.

A long bone consists of a shaft, or *diaphysis* (dy-af'i-
sis), and knobby ends, each of which is an *epiphysis*
(ee-pif'i-sis). The knobby epiphyses fit into appropriate
spots in adjoining bones, are coated with cartilage, and
in youngsters at least are separated from the bony parts
of the diaphysis by a stretch of more cartilage. Osteoblasts
in the bony portion of the diaphysis continually invade the
cartilage in the direction of the epiphyses, laying down
hydroxyapatite as they do so; and the cartilage continually
grows away from the shaft itself, pushing the epiphyses on
ahead. The net result is that the bone grows longer and
longer.

Finally, somewhere in the middle or late teens, the relentless layering of bone catches up with the epiphyses and
wipes out the cartilage between. The bones no longer

lengthen and the youngster reaches his or her adult height. One of the reasons why women are generally shorter than men is that this completion of the process takes place at a younger age in their case.

The Greek meaning of "diaphysis" is "growth between" and of "epiphysis" is "growth upon." The epiphysis, in other words, is a piece of bone growing upon another bone but separated by cartilage from it, while the diaphysis is the growth between the epiphyses at the two ends.

This complicated layering and delayering of mineral substance and this careful race of bone and cartilage cannot, as you can well imagine, be left to the bone itself. There must be some central directing force handling all the bones in order that the growth of each bone continue in due proportion with that of every other bone, and with the soft parts of the body, too. Such central control is exerted, in part, through the action of *growth hormone,* a chemical released into the bloodstream in tiny quantities by a small organ just under the brain called the *pituitary gland* (pi-tyoo'i-ter-ee). The presence of growth hormone keeps cartilage ahead in the race, so to speak.*

When something goes wrong with the supply of growth hormone, the results can be drastic. An undersupply will bring about a quick overrunning of the cartilage. Such a rapid ossification can put an end to growth in early childhood and the result is the circus midget. (Where the long bones are particularly affected, so that head and torso are about normal in size and the arms and legs remain stubby, the result is a "dwarf.") On the other hand, an oversupply of growth hormone can put cartilage so far ahead that a youngster may shoot up with unusual rapidity and continue to do so into adulthood. A circus giant results. Men have been definitely known to have reached heights of almost nine feet, and some midgets have been less than two feet high as adults.

Occasionally, there is an abnormal production of growth hormone after a normal complete ossification has taken place. In this case, further growth is induced at the only

---

* Except for passing mentions such as this one, I shall not discuss hormones and hormone action in this book. It will be more useful to leave such "chemical controls" to be dealt with along with the "electrical controls" of nerve, spine, and brain in the book I plan to write as a companion volume to this one.

places where growth remains possible, even under such stimulation—at the ends of the limbs and the tip of the lower jaw. Hands, feet, and jaw enlarge grotesquely, and this condition is known as *acromegaly* (ak'roh-meg'uh-lee; "large extremities" G).

Also involved in the production of bone is *vitamin D,* whose more formal name, *calciferol* (kal-sif'uh-role; "calcium-carrying") indicates its function. Children who for one reason or another happen to endure a shortage of vitamin D have bones that do not properly ossify. They remain soft and therefore deform under stress, giving rise to bowlegs and a curved spine. The skull may be soft and misshapen, a condition known as *craniotabes* (kray'nee-oh-tay'beez; "wasting of the skull" L). The disease, in general, is called *rickets* or *rachitis* (ra-kigh'tis; "spine" G, which, after all, is a portion of the body often affected). The effect of rickets is to be seen from the meaning of the adjective "rickety." Nowadays, with the widespread fortification of milk and bread with vitamin D, and the use of vitamin pills, rickets is not a serious threat, at least in the more developed portions of the world.

The vitamin D requirement of adults, once bone growth is over, is very low, and yet it may not be entirely zero. Mineral matter deposited in bone is not there permanently. It can always be mobilized by the body in case of need, so there must always be a mechanism to replace it when the need has passed. The lack of vitamin D may be one reason why adults sometimes suffer bone-softening, as mineral matter is removed from the bones and is not replaced. This condition is found in women more often than in men, particularly in the Orient. Usually it makes its appearance during pregnancy or lactation, when the mother's calcium supply is being stripped on behalf of the developing infant. This condition is called *osteomalacia* (os'tee-oh-muh-lay'-shee-uh; "bone softness" G) and has symptoms similar to those of rickets.

Infection of the bone marrow, sometimes a serious disease that has required surgery, is *osteomyelitis* (os-tee-oh-my-uh-ly'tis; "inflammation of the bone marrow" G). (The suffix "-itis" has come, by general agreement among medical men, to be used to signify "inflammation of." You can see at once, then, what such common words as "tonsillitis" and "appendicitis" must mean.)

TOOTH STRUCTURE

Like the bones, the teeth are built up about central channels containing nerves and blood vessels, so that they too incorporate living parts. There is but one such channel in each tooth and it is the *pulp* (which contains the nerve). It is the sensitive portion of the tooth, as is graphically portrayed by the common phrase "hitting the nerve."

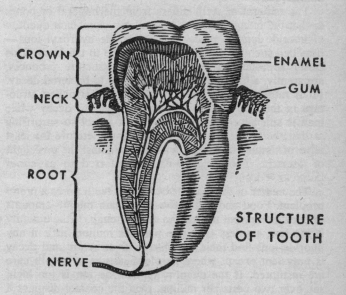

CROWN

NECK

ROOT

NERVE

ENAMEL

GUM

STRUCTURE
OF TOOTH

Immediately about the pulp and making up the bulk of the tooth is *dentine* ("tooth" L), which is much more mineralized than bone. Dentine is up to 70 per cent inorganic salt, in place of the 45 per cent that makes up bone. Dentine is therefore harder than bone and also less active. It exchanges material with the bloodstream to only a tenth of the extent that bone does. The ivory used for billiard balls and piano keys is an example of virtually pure dentine when derived from elephant tusks.

The dentine of that portion of the tooth below the gumline (the *root*) is surrounded by a thin layer of *cementum* which serves, as the name would indicate, to anchor the tooth to the jaw. Cementum is itself quite like bone in composition.

On the other hand, the dentine of the portion of the tooth above the gumline is overlaid by *enamel*. Whereas cementum is less mineralized than dentine, enamel is more so. Enamel is actually up to 98 per cent inorganic and is almost entirely inert. It is the hardest substance in the human body.

The mineral of teeth differs from the mineral of bones in that the former contains a small but significant quantity of fluoride ions in place of some of the hydroxyl ions— provided such fluoride ions are available to the body. This closer approach to the fluoroapatite structure produces, apparently, a tooth that is less amenable to bacterial decay. (It is rather ironical that the hardest and strongest structure in the body is the only one subject to decay while man is still alive—and yet perhaps it is not so surprising at that; being the most mineralized and therefore the least alive of tissues, it is most defenseless against the onslaught of bacteria. Tooth decay can be referred to as *caries* [kair′eez; "decay" L].)

The matter of the fluoride content of teeth poses a pretty problem. Food and water always contain minute amounts of fluoride ion, but not always quite enough. If the quantity is too low, say less than one part per million, little if any fluoride ions find their way into tooth structure and decay is prevalent except where heroic measures of mouth care are instituted. If the quantity of fluoride ion is too high, say over two parts per million, then the enamel displays a permanent yellow mottling that is not actually harmful but is certainly not pretty.

At about one part per million, the fluoride ion content reduces dental caries to one third its usual incidence (where no other change in oral hygiene is instituted) without any observable harm of any kind. (This last conclusion is based upon a quarter century of painstaking dental and medical research.) There is therefore a strong movement among the dental profession in favor of the addition of fluoride ion (*fluoridation*) to drinking-water supplies in order to bring the fluoride ion content up to the one-part-per-million level. It is estimated that dental bills could in this way be reduced by a billion dollars a year, with all the saving (not capable of being measured in money) in fear and pain that would imply.

Unfortunately, the reduction in caries would apply only to children during those ages when their teeth are forming and incorporating fluoride ions. Adults, with fully formed teeth, will no longer incorporate fluoride, but at least the new generation would be started off on the right foot.

BONE MOVEMENT

The skeleton is not merely a framework for the body, it is a movable framework. Since the bones themselves are rigid, the only possibility of motion comes at those points where two bones join. These points of joining are called, very obviously, *joints*. More ornately, they can also be called *articulations* ("to join" L). The existence of a joint does not by itself necessarily imply mobility. Some bones, such as those of the skull and of the hipbone, fuse together, as I explained earlier, into what is essentially a single structure, with no movement of any kind possible at the lines of joining.

Other joints allow a mere gliding movement, and not much of that. Examples include the joints between the vertebrae and those between the ribs and the thoracic vertebrae. These allow the limited motions involved in the bending of the back or the lifting of the rib cage during breathing. The small bones of the wrist and ankle can also glide one against the other. The joint motions with which we are most familiar, however, involve sharp and extreme changes in the relative positions of neighboring bones. These are most marked in the limbs, as when you bend your arm at the elbow or your leg at the knee. The motion there is virtually through an angle of 180 degrees.

With one bone moving so against another, one important concern (as it would be in an equivalent man-made apparatus) is to reduce friction. For this reason the portions of the bone that meet are lined with a smooth layer of cartilage. The bones are also held together by a capsule (*synovial capsule*) of connective tissue that encloses the joint and secretes a viscous liquid containing hyaluronic acid. The joints slide easily against this lubricating layer of *synovial fluid* (si-noh'vee-ul; "egg white," which it resembles in its smooth thickness). Joints about which more or less free movement is possible are, for this reason, referred to as *synovial joints*.

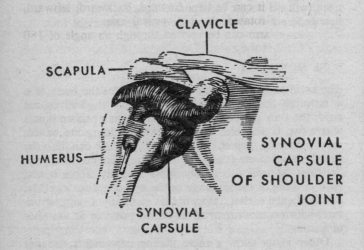

CLAVICLE

SCAPULA

HUMERUS

SYNOVIAL
CAPSULE

SYNOVIAL
CAPSULE
OF SHOULDER
JOINT

The type of movement allowed at a particular synovial joint depends upon its structure. As a consequence, motion is sometimes possible in one plane only, back and forth like a door upon its hinges, so that such a joint is a *hinge joint*. One example is at the elbow where the proximal epiphysis of the ulna just fits between the two epiphyses at the distal end of the humerus. It can therefore move back and forth but not from side to side.

The knee joint is another hinge joint. So are the joints between the first and second phalanges of fingers and thumb, and between the second and third phalanges of the fingers. (This also holds true for the analogous joints in the toes, so that in the limbs there are 40 hinge joints in all.)

Some joints can allow motion in each of two axes. For instance, you cannot only bend your toes; you can also spread them apart. This is even truer of the fingers.

The lower jaw can be moved up and down and is mostly a hinge joint, but it can be moved somewhat from side to side and the normal action of chewing involves a rotary motion rather than a mere clashing of teeth. (Watch a cow chewing its cud if you want to see such rotary motion in slow and stately fashion.) The head itself is even more freely movable about its connection with the vertebral column, since it can be bent forward, backward, leftward, rightward, or rotated about a vertical axis.

The lower arm can be rotated through an angle of 180 degrees so that the palm of the hand can face either down or up, without motion at either elbow or shoulder. This is made possible by the manner in which the proximal epiphysis of the radius fits into a depression in the ulna. Within that depression the radius can pivot. If you hold out your arm before you, palm up, radius and ulna are parallel; turn the arm palm-down and the radius pivots and crosses the ulna. (The foot is much less versatile than the arm in this respect.)

When the epiphysis of one bone fits into a cup-shaped socket in another, you have a *ball-and-socket joint*. The most obvious case is that of the femur fitting into the acetabulum of the hipbone. This allows the freest possible motion so that the leg can be thrown into almost any position, especially with practice, thus making ballet dancing the graceful thing it is.

A similar ball-and-socket joint between the humerus and the scapula involves even freer motion, since the socket is shallower in this case than in that of the hip. You can turn your arm through a complete circle at the shoulder and this joint is by all odds the most maneuverable in the body. (Watch a baseball pitcher in a complicated windup.) This is a good thing too, considering that the possession of arms and hands capable of almost infinitely versatile manipulation is one of the factors that is the making of the human being.

Violent movement about a joint can succeed in moving one bone out of alignment with another (*dislocation*), which results in making motion at that point impossible and attempted motion extremely painful. A ball-and-socket joint is more easily dislocated than any other kind and the shallow shoulder joint is most easily dislocated of all, with the hip joint next. The elbow is sometimes dislocated, as are the various phalanges. One misadventure that is almost humorous (to everyone but the victim) is the dislocation of the lower jaw as the result of a too vigorous yawn.

To prevent dislocations, as far as possible, it is not enough to depend on the synovial membranes or the pressure of surrounding muscles to keep the joints from coming unhooked. Neighboring bones at synovial joints are held together by strands of tough connective tissue called *ligaments* ("to tie" L). Ligaments help restrict motion at the joints to certain reasonable limits. However, those limits can be exceeded under extreme conditions so that ligaments can be torn, with or without dislocation of the joint itself. Such *sprains* occur most frequently at the wrist or ankle. The resultant pain and swelling are familiar to all of us for there are few, if any, who can avoid sprains altogether.

Ligaments may be either white or yellow. White ligaments are composed chiefly of collagen and are not elastic. Yellow ligaments contain elastin and are, therefore, elastic. The former are common but the latter are rare and are found only in the neck in man.

Strong white ligaments bind the bones of the foot in such a way as to curve them into an arch. It is the springy ligament that cushions the shocks of locomotion, and it is the giving of those ligaments that results in flat feet.

Despite all precautions, moving parts are a particular prey to disorder and are as vulnerable in the body as they would be in manufactured products. The knee (perhaps the most vulnerable joint in the body despite the added protection of the patella) may accumulate synovial fluid after injury; this condition is popularly known as "water on the knee." Or the synovial pouch, the membrane of connective tissue enclosing the joint, may grow inflamed and painful. This can happen when steady pressure is placed on the knee, as traditionally among housemaids, who were forever scrubbing floors (in the old days when floors were scrubbed and housemaids existed), so that the disorder came to be called "housemaid's knee." The synovial pouch is called a *bursa* ("purse" L, because it seems to enclose the joint as a purse would enclose its contents). Inflammation of the synovial pouch may therefore be called *bursitis*. It strikes often in the shoulder.

Any inflammation of the joints, for whatever reason, is a type of *arthritis* ("joint inflammation" G). The most serious and widespread variety is *rheumatoid arthritis,* the cause of which is unknown but which can attack anyone at any age, although it most often strikes between the ages of 30 and 45. (It is called "rheumatoid" because it has the symptoms once associated with what was called "rheumatism"; that is, pain at the joints.) In addition to the general pain and misery it brings on, the disease in its extreme manifestations may destroy the structure of a joint or even immobilize it permanently by fiber formation and the deposition of mineral matter. In this way, rheumatoid arthritis can, in effect, produce a bedridden patient.

4

OUR MUSCLES

LIVING MOTION

Although the skeleton considered by itself clearly provides the possibility of motion—after all, it is jointed—it does not and cannot move of itself. The skeleton, when used as a fright-object for children in stories and in animated cartoons, is most terrifying when its lank bones stir in pursuit and its skinny arms stretch out in menace. However, it takes no more than the tiniest advance in sophistication to know that bones, even fresh bones with all their cells intact and alive, could no more move of their own accord than a plaster cast of those same bones could. For motion we must look elsewhere; if there is one characteristic we strongly associate with life, it is that of voluntary movement.

We associate such movement with animal life particularly, for a casual inspection would lead us to suppose that plants do not move except where motion is forced upon them by wind or water current. This is not entirely correct, of course. Plant stems will turn slowly in the direction of light and away from the pull of gravity, while plant roots will turn slowly in the direction of water and toward the pull of gravity. This slow motion depends, apparently, upon differential growth. That is, the cells on one side of the stem or root divide rapidly and the cells on the other side divide slowly so that the structure bends in the direction of the nongrowing side. If light or humidity inhibits growth on the side of the structure it reaches, that structure will turn toward the light or water.

For more rapid motions, in response to touch or light, plants make use of water turgor, which means that certain cavities at the base of petals can be filled with fluid under pressure. When those bases stiffen, the petals are pulled open. When the cavities empty, the petals grow limp and close. This is a primitive device but animals are no stranger to it. The human body contains portions which ordinarily flaccid, can swell and become rigid when sponge-like cavities fill with blood under pressure. The best-known example is, of course, the male penis.

Nevertheless, none of this is what we usually think of when we think of life in motion. We think of antelopes, horses, cheetahs, ostriches (and, in a bumbling way, we ourselves) racing across the ground; we think of the flying of birds, bats, and insects; of the slithering of snakes; of the swimming of fish and porpoises; of the burrowing of moles; and so on. (Yet there are animals, such as clams and coral, which throughout most of their lives are scarcely less immovable than plants are.)

If we are to discover the mechanism of motion, we must turn to the cell, which is the biological unit of life. And we find that all cells—those of men, eagles, clams, and sycamore trees alike—show a capacity for internal movement. The protoplasm within the cell constantly circulates in a definite pattern. This is sometimes called *protoplasmic streaming* and, sometimes, *cyclosis* (sy-kloh'sis; "circulation" G).

The value of cyclosis for any cell is that of keeping its contents well distributed, of making sure that different parts spend a fair share of the time near the membrane where material can be picked up from the outside world or discharged into it. Also, material could be transported, by cyclosis, between the membrane and vital cell structures more or less permanently placed in the interior.

Such streaming can be modified so as to result in the bodily movement of a cell. The protoplasm of a cell can exist in one of two states: as a stiffish semisolid called a *gel* (short for "gelatin"), a protein that when properly mixed with water offers the best-known example of such a state; or as a freely moving fluid called a *sol* (short for "solution"). The balance between the two states is a delicate one, a small change sufficing to switch an area of protoplasm from gel to sol or from sol to gel.

Imagine the protoplasm along the central axis of a cell to be in the sol form, and the portion surrounding it gel. If the gel at the rear were somehow to contract, it would squeeze the sol forward like toothpaste out of a tube. The forward portion of the cell would bulge outward.

As the sol streamed forward, it would turn to gel along the walls, while some of the gel in the rear would turn to sol and in its turn be pushed forward. In this way the capacity for internal motion is channeled into movement forward and the entire cell creeps.

This form of movement has been most thoroughly studied in a one-celled animal called the *amoeba* (uh-mee′buh; "change" G) and is therefore called *amoeboid movement*. The very name of the creature arises from its mode of locomotion—as the cell bellies forward in this direction or that, forming *pseudopods* (syoo′doh-podz; "false feet" G), its shape is constantly changing.

Other one-celled creatures have, over the ages, developed specialized attachments that make more rapid movement possible. These are microscopic hairlike structures to lash the water and move the cell. The attachments may be few and comparatively long, in which case they are called *flagella* (fla-jell′uh; "whip" L), the singular being *flagellum*. The attachments may also be many and short, in which case they are *cilia* (sil′ee-uh; "eyelash" L), the singular being *cilium*.

Although these forms of motion may strike us as being adapted particularly for primitive cells, the lordly multicellular creatures that have developed over the eons have not abandoned them. Consider man, for instance. There are cells in our blood that exhibit amoeboid movement, crawling about within us by the old gel-sol interchange. The human sperm cell makes its way toward the egg cell by means of the lashing of a single flagellum. (The fact that it is called a "tail" doesn't make it less a flagellum.) Finally, there are ciliated cells in the respiratory system and in the female reproductive system, the whipping cilia of the first serving to brush foreign matter out of the lungs, and those of the latter serving to brush the egg cell from the ovary to the uterus.

It would seem that the flowing-forward of sol in amoeboid movement is the result of a contraction of the gel in the rear. It is also contraction that may be the cause of the motion of cilia and flagella. Both cilia and flagella are composed of eleven fine filaments, nine of which form a circle about a central pair. One theory as to the cause of their motion is that the filaments first on one side of the center and then on the other contract, bending the whole structure back and forth.

The contraction involved in amoeboid movement and in the whipping of cilia is only conjecture so far, but it seems quite reasonable that this ability to contract a part of itself is a very basic property of the animal cell. After all, quite early in the evolution of multicellular life, certain cells were developed which dedicated their lives, so to speak, to contraction. Their contraction is visible and unmistakable, and it is reasonable to suppose that such a specialty is not created out of nothing, but that it represents the extension and exaggeration of a property already existing, in more dilute form, in cells generally.

MUSCLE CONTRACTION

Cells specializing in contraction make up those portions of the body we call *muscles,* and the individual cells are, therefore, *muscle cells.* The word "muscle," according to one theory, comes from a Latin word meaning "little mouse," because a man can make his muscles ripple in such a fashion as to make it look as though a little mouse were running about under his skin. This seems a bit fanciful; another theory, which I like better, has the word arising from a Greek expression meaning "to enclose," because layers of muscle enclose the body.

There are several types of muscle tissue in the human body, which can be distinguished in a number of ways. Under the microscope, for instance, certain muscles are seen to consist of fibers that have a striped, or striated, appearance, with alternate bands of lighter and darker material. These are *striated muscles*. Another type of muscle, lacking these bands, is the unstriated muscles, or *smooth muscles*. (There is also a type of muscle, not quite like either, that makes up the structure of the heart, but this I will consider in a later chapter.)

It is striated muscle that has been most carefully studied in fine detail. Under polarized light, the darker stripes refract light in different ways, according to the direction of the light beam. Whenever the property of some structure varies with direction, it is said to be *anisotropic* (an-eye'soh-trop'ik; "turning unequally" G), so the darker stripes are *anisotropic bands* or, simply, *A bands*. The lighter stripes do not change properties with direction and are *isotropic* ("turning equally" G), so that they make up the *I bands*. The A band is divided in two by a thin line called the *H disc* (this being short for *Hensen's disc*, named for Victor Hensen, a German anatomist of the 19th century). Down the middle of each I band is a dark line called the *Z line*. Using the electron microscope, it would seem that the A band consists of a series of wide filaments, and the I band consists of a series of thin filaments anchored to the Z line. When the muscle fiber is relaxed, the overlapping is not complete and adjacent I band filaments do not meet. It is the gap between the I band filaments that make up the H disc.

Now the stage is set for movement. A nerve impulse reaches the fiber and this sets into action a series of chemical changes that, among other things, liberate energy. Changes visible under the electron microscope make plain what happens as a result. The I band filaments inch toward each other, dragging the Z line anchor with them. The I bands meet each other between the A band filaments, wiping out the H disc.

The A band filaments remain essentially unchanged so that, to put it as simply as possible, the muscle fiber in relaxation is the full length of the dark bands plus light bands. In contraction, however, it is the length of the dark bands only.

What makes the I band filaments move as they do? This is not quite settled, but the most reasonable conjecture is that there are tiny extensions of the A band filaments

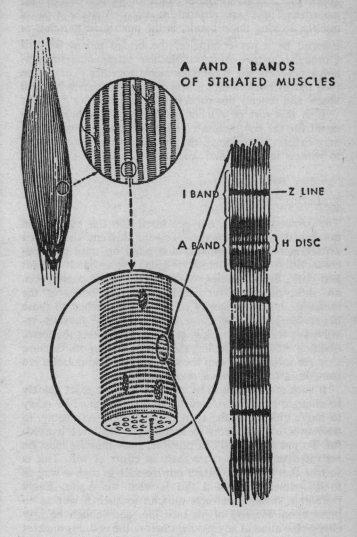

A AND I BANDS
OF STRIATED MUSCLES

I BAND — Z LINE

A BAND } H DISC

which can connect at specific places with the I band filaments. The connection is at a backward angle and under the inflow of energy arising from the chemical changes induced by a nerve impulse, the bridge moves forward (through changes in intermolecular attractions, presumably) dragging the I band filaments together. The bridge then moves backward to a new connection, moves forward again, and repeats this process over and over.

It is this sort of ratchet arrangement which may account for the contraction of all muscles and consequently may explain virtually all the motions we are accustomed to witness in the animal world.

STRIATED MUSCLE

The striated muscles are organized for the purpose of quick and hard contraction. Toward this end the ratchet arrangements are multiplied and piled up until they become visible to the microscope-aided eye as bands of light and dark. Such muscles, designed for quick contraction, are most necessary for manipulating the bones of the skeleton, and the majority of them are attached to those bones for just that purpose. Those are the muscles involved in walking, jumping, grasping; in twisting the torso, nodding the head, expelling the breath, and so on. Because they are so involved with the skeleton, striated muscles are sometimes called *skeletal muscles*.

If such muscles are to be useful, they must react quickly to changes in the environment. A creature must be ready to move quickly at the sight of food or of an enemy, to maneuver without delay to accomplish whatever is needful. To put it as briefly as possible, such muscles must be activated through no more than an effort of will. This is so true that we can contract our muscles in such a way as to do ourselves harm if this is what we desire. Every deliberate suicide contracts muscles in such a way as to bring about the end of his own life, and though he may change his mind at any moment before the end, his muscles will not refuse to obey him if he doesn't change his mind. For this reason, striated muscles are also called *voluntary muscles*.

The smooth muscles, on the other hand, are poorer in the contracting filaments (poor enough to be unstriated, but, although they have not been well studied, it is assumed that they contract through the same mechanism striated muscles do). As a result, smooth muscles contract slowly and are restricted to organs where quick motion is not so vitally required. They are found making up the walls of the internal organs, such as those of the blood vessels, and the digestive tract. Since the internal organs are referred to as *viscera,* the smooth muscles are sometimes called *visceral muscles.*

The visceral muscles react relatively slowly to changes within the body and do so without the intervention of the will. The walls of the blood vessels contract or expand to response to certain chemicals in the blood or in response to the effects of temperature, but we cannot of ourselves deliberately cause them to do either in the same effortless way in which we lift our arm or lick our lips. For this reason, smooth muscles are also called *involuntary muscles.*

On a cellular level, the differences are also meaningful. The individual cells of smooth muscle are cigar-shaped (they are usually described as spindle-shaped, but there are few people now alive who are intimately acquainted with spindles or know what they look like) and have but one nucleus. The striated muscles, on the other hand, are made up of cylindrical structures that differ from ordinary cells in having a number of nuclei placed here and there, as though a number of cells have been combined in order that the total force of contraction might be the greater and the better coordinated.

Muscle fibers contract on the "all-or-none principle." That is, a stimulus (which can be an electric current, the pressure of touch, the action of heat or of certain chemicals) either applied directly to the muscle or to the nerve that leads to it may be so weak that the muscle is not affected at all. If the stimulus is strengthened, a point is reached where the fiber responds, and it does this in the one way it knows—the complete contraction of which it is at that moment capable. There is no intermediate stimulus that will bring about an intermediate contraction. It is all or none.

This seems to be at variance with actual experience,

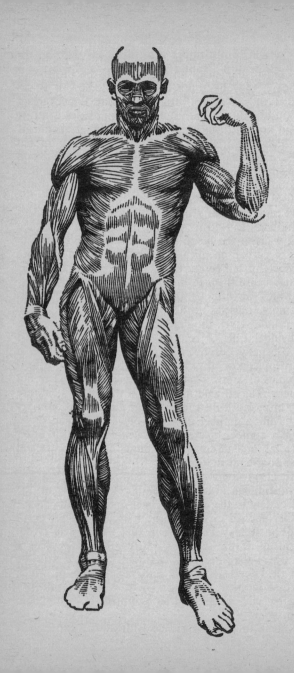

for the muscle of our upper arm, to name the one with which we are most familiar (it is the one children contract to "make a muscle"), can be made to contract to any degree, from the barest twitch that just moves the lower arm a hairbreadth to a rapid and maximal flexure at the elbow.

This is no paradox but arises from the fact that an actual muscle is made up of many fibers which, to a certain extent, have the capacity for independent action. If only a few fibers of the muscle are activated, those few will contract all the way, but the net effect of those few "alls" against the background of the very many "nones" will result in only a slight total flexure. As more fibers respond to the stimulus, the overall contraction grows gradually greater.

The response of a fiber of striated muscle to a stimulus is quick and brief, lasting only a fraction of a second; only 1/40 of a second in some cases. This is followed by a quick relaxation. A single quick stimulus of this sort is called a "twitch" and does not normally occur in the body. More usual is the slower contraction of a whole muscle, or even its continued contraction over some reasonable period of time, this being produced and maintained by various fibers taking their turn at twitching. No one fiber is contracted for very long, but the muscle as a whole can be contracted for quite a while. It is, however, possible to induce a single twitch in a body muscle. The best example is the sharp blow under the knee which gives the familiar response of the "knee jerk."

It is possible to stimulate an isolated fiber so that it will respond by a series of twitches. If the stimuli are close enough together, each twitch will build on the one before it. That is, after one twitch, a muscle will not have time to relax completely before another twitch is called for. The second twitch, beginning at a state of greater tension, achieves a greater force, and the third twitch, hurrying on the heels of the second, a still greater one. By the time stimuli reach 50 per second the fiber is in a continuous tight contraction. This is called *tetany* (tet'uh-nee; "stretched" G—even though the muscle is contracted rather than stretched).

Tetany does not take place in the muscles of a well-organized body, but bodies are not always well organized. Even those capacities that are subject to our will can get out of control. Our body is a chemical machine and it remains obedient to us only when its chemistry is in order. As an example, the proper transmission of the nerve impulse to the muscle depends upon the proper concentration of certain metallic ions in the blood. One of these is calcium ion. If for any reason the calcium ion concentration sinks below a certain minimal level, the nerve starts to deliver rapid impulses and muscles respond by going into tetany. Death will follow if the condition is not relieved.

The calcium ion concentration in blood is well controlled by the body and is not likely to go wrong. Danger threatens more often from without. There is a bacterium called *Clostridium tetani* (klos-trid'ee-um; "spindle-shaped" G), which is all too common and can enter the body through any wound. It produces a *toxin* (anything poisonous may be called a "toxin" [from the Greek word for "arrow"] because of the old habit of poisoning arrow tips), which in turn causes muscles to go into tetany, with usually fatal results. The disease is called *tetanus* or, graphically and rather horribly, "lockjaw," because the earliest muscles affected are those of the jaws, which contract and lock firmly.

Tetanus was once one of the more deadly dangers that accompanied wounds that might not otherwise have been of much concern. Fortunately it has been found possible to treat the tetanus toxin chemically so as to alter it slightly and produce a *toxoid*. This toxoid will not induce tetany but will encourage the body to produce substances (*antibodies*) that will not only neutralize the molecules of toxoid but will also neutralize any molecules of the toxin itself that enter the body. In this way a series of well-spaced "shots" of tetanus toxoid will produce immunity to the disease. During World War II American soldiers were constantly being punctured to produce immunity to tetanus, and other diseases too, and although the procedure was the subject of many heartfelt jibes on the part of the victims of the needle, tetanus virtually disappeared. This is worth a needle.

Symptoms similar to that of tetanus, and with the same

fatal result, are brought about by the drug *strychnine* (strik'nin; "deadly nightshade" G, the plant from which it is obtained). Exactly how either strychnine or the tetanus toxin produces its effect is not known.

Muscles will also stiffen and grow hard during the period from 12 to 36 hours after death, this being the *rigor mortis* ("death-stiffening" L) made famous by murder mysteries (which also did their bit in familiarizing the general public with strychnine). Rigor mortis is not due to a tetany but to the precipitation of ordinarily soluble proteins in muscle tissue. The effect is rather like that of hard-boiling an egg. As decomposition sets in, rigor mortis vanishes.

The price paid by skeletal muscle for its gifts of speed and force is fatigue. The chemical changes induced by nerve stimulation of muscle inevitably deplete the relatively small supply of energy-producing substances in the muscle cells. In one way or another these substances must be replaced as quickly as they are consumed, and this is usually done by means of other chemical reactions involving oxygen molecules brought to the muscle cells by the bloodstream, which in turn picks up those molecules at the lungs.

The visceral muscles, working slowly and in response to orderly changes in the internal environment, maintain a pace that matches their oxygen supply. They keep their energy-producing substances at the necessary level at all times and are immune to fatigue.

The skeletal muscles, however, are subject to levels of activity much higher than normal over extended periods of time. It might seem necessary, then, to see that the blood supply is correspondingly greater at all times "just in case," and yet it would be inefficient to design muscles for continuous work at ditch-digging level when most life activities are considerably less intense. Still, it is sometimes necessary to dig ditches or chop wood or run at top speed to escape a danger. For that matter, it sometimes seems desirable to play several sets of tennis. On such occasions, the body can adapt itself to the greater need for oxygen. Hard work or hard play results in an increase in both the rate and depth of breathing (we pant), and in the rapidity of the heartbeat (it pounds) and in the capacity of the blood vessels feeding the muscles (we flush). And yet survival may depend upon finding a way, at least temporarily, to go beyond all that lungs and heart and blood can do.

The muscle accomplishes this by obtaining a supply of energy (a limited supply, to be sure) at the expense of chemical changes that do not involve oxygen. In these chemical changes, a compound known as *lactic acid* is produced.* This dodge, which allows the muscle a possibly life-saving extra push, is not without its price. As lactic acid accumulates, it becomes more difficult for the muscle to contract, and the sensation we experience is that of fatigue. We slow down perforce, and eventually, even with life at stake, we must stop in utter exhaustion.

When fatigue stops us, we cannot recover entirely until the lactic acid is removed from the muscle. This requires oxygen—all the oxygen that would have been used up if the total effort expended had been at a rate slow enough for the oxygen-supplying capacity of the body to have kept up. We have incurred an "oxygen debt" that must be paid off. It is done by supplying the muscle with oxygen at top rate until the lactic acid is partially burned away and partially rebuilt into the original energy-producing substances. For this reason, we continue to pant and flush and our heart continues to pound for quite an interval after we have ceased our activity and crumpled exhausted onto the ground or into a chair.

Just as striated muscles are capable of harder and more intense work than visceral muscles and are therefore also capable of fatigue, so there are gradations among the striated muscles themselves. Those muscles designed for particularly quick and hard contraction are more easily fatigued than those not so designed. In general, the particularly quick and particularly easily fatigued muscles are paler in color than the slow but more enduring ones. The division is most plainly visible to us when we eat chicken or turkey. The breast muscles, intended for operating the wings and therefore designed for hard work, make up the "white meat." The leg muscles, intended for less

---

* See my book *Life and Energy* for the details.

intense work over longer-sustained intervals, make up the "dark meat."

In man, too, there are darker and lighter muscles fibers. When a man stands, the large muscles of the back are continually twitching him this way and that in order to maintain him in balance. (A two-legged stance is not very stable; when a man gets a trifle drunk, so that his muscular coordination is thrown off, it becomes all too easy for him to fall down.) These back muscles are of the dark slow-to-fatigue variety, and even though we eventually tire of standing we tire of digging much more quickly, let us say, where the quick-to-fatigue arm muscles come into play.

There is a long-term adaptation to the intense use of muscles which is designed to decrease the effect of fatigue. The muscle itself, under the stimulus of long continued use at strenuous levels (whether because of the necessity of physical labor or the whim of physical exercise), increases in size. This is called *hypertrophy* (hy-per'troh-fee; "growth beyond" G). The lumberjack or athlete has larger muscles than the storekeeper or file clerk and in those larger muscles can store greater supplies of energy-producing substances and find more room for lactic acid. Lungs of greater capacity and a heart of greater pumping force supply the large muscles with extra quantities of oxygen, and the result is to make possible greater force over more extended periods with less fatigue.

On the other hand, though lack of exercise keeps a muscle relatively weak and small, under ordinary conditions there is no danger that a muscle will lose its ability to function at reasonable levels, however sedentary a man's life. For one thing, the normal muscle maintains a weak contraction even when the body seems to be relaxing. This is called *muscle tone* and, in a sense, it means that we are constantly exercising.

Muscle tone serves to keep the individual muscles in greater readiness for contraction at short notice. The muscles begin with a headstart, so to speak. People under nervous tension usually have a more intense muscle tone and require a smaller stimulus to set them off. For that reason, they twitch and are "jumpy." Nevertheless, the principle remains valuable, even if it can be overdone.

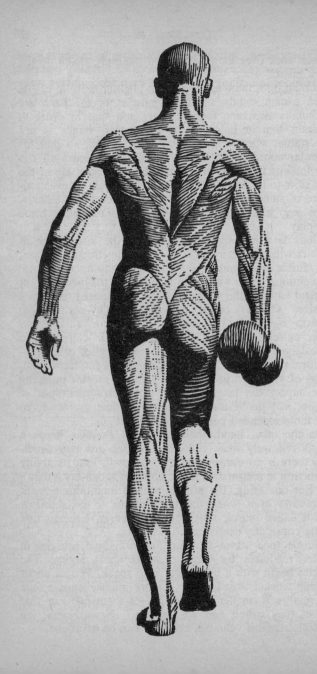

During sleep, muscle tone diminishes to minimal levels and muscles can experience periods of virtually complete relaxation; something they cannot experience during wakefulness. This imposed relaxation of muscle is undoubtedly one of the vital functions of sleep. It is one of the reasons why sleep is not only pleasant but is actually more necessary than food. (Sleeplessness is far worse torture than hunger, and a man will die for lack of sleep sooner than he will starve for lack of food.)

If muscle tone is removed permanently, as by cutting the nerve that leads to the muscle, then the muscle does indeed waste away. It undergoes *atrophy* (at'roh-fee;" no growth" G). If a nerve-destroying disease such as poliomyelitis paralyzes the legs, the leg muscles wither even though the rest of the body remains strong and well-developed.

The visceral muscles, on the other hand, retain their tone and therefore their usefulness even without nerve stimulation, so that they are not affected by polio. In fact, the greatest immediate danger of the disease is that it will paralyze the muscles that act to move the rib cage up and down. These muscles are striated and make breathing possible. When these are paralyzed, an "iron lung" becomes necessary; when a patient is in one, changes in air pressure will do for him what the muscles of the chest do for one not so unfortunate.

TENDONS

The muscles most familiar to use are those of the arms and legs, since we use our limbs freely and the muscles involved change shape visibly in the course of limb movements. Consider the muscle of the upper arm: it is thick in the middle (the "belly" of the muscle) and narrows as it approaches the bone. It stretches across the elbow joint and is attached to each of the two bones (the humerus and the radius, in this case) meeting at that joint. When this muscle, or any similar one, contracts, one of the bones to which it is joined remains virtually stationary as a result of the steadying action of other muscles ("fixation muscles") joined to that bone. The point of junction of the muscle with the stationary bone is the *origin* of that muscle. When the muscle contracts, then, it is the second bone that moves, pivoting about the joint to approach the first bone. The attachment of the muscle to the bone that moves is its *insertion*.

There can be more than one origin, as is actually the case with the muscle of the upper arm, where one origin is at the upper end of the humerus near the shoulder joint, and a second origin reaches above to the scapula. This muscle is an example of a *biceps* ("two heads" L) and is sometimes called just that. Actually it is not the only muscle with two origins so that its proper name is *biceps brachii* (by′seps bray′kee-eye; "two-heads of the arm" L). The insertion of the biceps brachii is at the upper end of the radius near the elbow. When the biceps brachii contracts, the humerus remains motionless and the radius moves toward it. The arm bends at the elbow in the familiar "making a muscle" gesture, and the "muscle" that is made is the contracted, thickened, and hardened belly of the biceps brachii.

Toward the point of insertion, a muscle usually narrows down to a strong tough cord of connective tissue, formed through a combination of the thin sheath of connective tissue surrounding each muscle fiber. This cord, attaching the muscle to the bone, is a *sinew* (AS) or *tendon* ("to stretch" G; because it stretches tightly from muscle to bone). Tendons are surprisingly strong, and in laboratory testing they have withstood a pull of as much as 9 tons per square inch before breaking. It can happen that in a violent muscular spasm the bone to which the muscle is attached will break sooner than the tendon connecting the muscle to the bone. To be specific: you can feel the tendon connecting the biceps brachii to the radius if you bend your arm into a right angle at the elbow and put you fingers at the inner angle of the joint.

Tendons serve to concentrate the full force of the muscle upon one spot on the bone. Naturally there are times when a group of bones must be moved and then the tendon must broaden and flatten out. For example, there is a muscle called *palmaris longus* (pal-may′ris long′gus; "long muscle of the palm" L) which has its origin at the humerus of the upper arm and its insertion at the *palmar fascia.* The term fascia (fash'-ee-uh; "band" L) refers to sheets of fibrous tissue encircling the body under the skin and enclosing muscles and groups of muscles. The palmar fascia would be the section of that fibrous tissue which underlays the skin of the palm.

The action of the palmaris longus is to flex the wrist. If you hold your hand before you, palm upward, you will be making use of that muscle when you bend the wrist so that the palm faces you. If this muscle had an ordinary tendon attached at some particular point at the palmar fascia, tension would be greatest at that point and the relatively broad palm would be under uneven stress. For that reason, the tendon expands into a widening flat sheet that attaches across the width of the palm. Such a wide, flat tendon is called an *aponeurosis* (ap'oh-nyoo-roh'sis; "from a tendon," because before the point at which it begins to widen the tendon is of the ordinary shape).

Tendons also serve to allow action at a distance, so to speak, when it is impractical to have muscles themselves on the spot. Thus, the usefulness of the fingers lies in their slim maneuverability, and if you feel them they seem little more than skin and bone. If

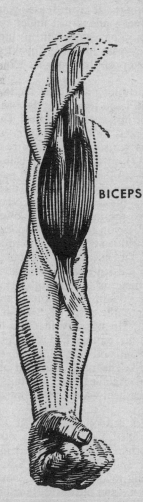

BICEPS

muscles were present in their structure, they would have to be built more thickly and softly and would lose much of their value. The muscles that do move them lie in the lower arm and palm. These possess tendons which run down the length of the fingers. The fingers are then flexed on the same principle that a puppet-master maneuvers his marionettes: by tightening strings. If you tense your fingers, claw fashion, you can feel tendons just under the skin of the back of your hands. You can see them, too, sometimes, especially where they pass over the knuckles. Similar tendons run across the top of the foot and along the toes.

As an example of a particularly long and stout tendon, consider the one attached to the large muscle in the back of the lower leg; you can feel it just below the knee. The lower leg bellies out at that point, and, as a matter of fact, the name of the muscle is *gastrocnemius* (gas'trok-nee'mee-us; "belly of the leg" G). It is commonly called the "calf muscle."

The muscle has two heads, both of them having their origin at the femur just above the knee. In the other direction it reaches down only to the middle of the lower leg (as you can see for yourself, if you tense it and look). Below is a long stout tendon with its point of insertion at the calcaneus, the bone of the heel. This tendon is called *tendo calcaneus* (ten'doh kal-kay'nee-us; "heel tendon" L) but a more common name is the "Achilles tendon." In Greek mythology, as you probably know, the warrior Achilles was dipped into the river Styx by his mother when he was an infant. He was invulnerable to weapons as a result. However, his mother had held him by one heel as she immersed him and had forgotten to dip the heel separately afterward. The result was that his heel remained unprotected and Achilles was eventually killed by a poisoned arrow in the heel—and left his name upon the tendon.

The existence of the Achilles tendon means that although the calf is powerfully muscled, the shin, ankles, and feet are slim. In the case of certain animals designed for running, such as horses, deer, and antelope, the lower portion of the leg (corresponding to our ankle and foot) is greatly lengthened. That section, as in ourselves, remains without large muscles. If you look at a horse you will see that powerful muscles are bunched at the top of the limb near the torso, whereas the legs themselves, equipped with tendons, are slim. This is important in connection with their ability to develop speed—a small contraction of the distant muscle will swing the thin, light leg through a large arc.

Another term for the tendo calcaneus of such animals is "hamstring," since it is the string (tendon) that connects the lower portion of the limb with the uppermost portion, or "ham." The action of those upper muscles is essential to walking, and to cut the hamstring of an animal is to cripple it.

In man, the tendo calcaneus connects the heel to the calf of the leg and not to the ham (or back of the thigh). You can, nevertheless, feel two tendons behind your knee, one on either side, which represent the insertions into the tibia and fibula of several muscles running down the back of the thigh. These muscles bend the leg at the

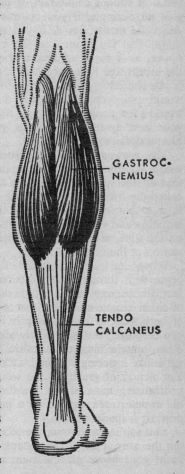

GASTROC-
NEMIUS

TENDO
CALCANEUS

knee and it is these tendons that represent the hamstrings in man.

MUSCLES IN ACTION

The one function of a muscle is to contract. When it has finished contracting, it will do nothing further but simply relax. It cannot counteract or undo its earlier action. To give an example, if you contract your biceps, your arm will bend at the elbow. If you arrange your arm in such a way that the lower arm lies above and rests upon the upper, as a result of biceps contraction, and then relax the biceps, the lower arm will continue to lie there limply. If you shifted your upper arm properly, the force of gravity would pull your lower arm downward and your arm would then dangle, extended. But how could you extend your arm against gravity (as you can) if relaxing the biceps doesn't do it?

The answer lies in the existence of a second muscle whose contraction serves to produce a motion contrary to that of the biceps. Opposite the biceps is a second muscle called the *triceps brachii* ("three-headed arm muscle" L). It is so named because it has three origins, two on the humerus and one on the scapula. Its point of insertion is on the ulna, on the side opposite that of the point of insertion of the biceps.

When the triceps contracts, the lower arm, if bent to begin with through biceps action, is straightened out. In fact, the bending of the arm is not the result of the action of either muscle on its own but is the well integrated action of both muscles simultaneously. The force of one muscle can be played out very slowly and smoothly by having the other muscle almost neutralize its force, but not quite. (It is like lowering a piano by a rope over a pulley system. The action of gravity alone would be disastrous, but by neutralizing most of the gravitational attraction through the counterforce applied by a man or group of men, the descent is made smooth.)

You can even contract the biceps strongly without moving the arm at all, if the triceps matches the pull evenly. Hold your lower arm at right angles to the upper and harden the biceps without moving your arm. You will feel the triceps on the other side of the humerus harden as well.

Muscles are always arranged at least in pairs, and usually in even more complex coordinating groups, so that one muscle or group of muscles will balance another. If there is a series of tendons acting to bend the fingers, a companion series on the other side of the fingers can act to straighten them.

Sometimes we can rely on gravity to undo the work of a muscle. One of the functions of the large gastrocnemius of the calf is to raise us on tiptoe. No equally large muscle on the other side of the tibia is required to bring us down from our tiptoe position. Gravity will serve the purpose. It is for this reason that the front part of the leg is relatively unmuscled and we can feel the tibia just under the skin all the way down.

The bones and the muscles attached to them as a lever system. The simplest form of a lever is the seesaw, in which the fulcrum is at the center of the lever and two approximately equal weights (in the form, usually, of little boys or girls) are at either end. The only purpose is to change the direction of the force, so

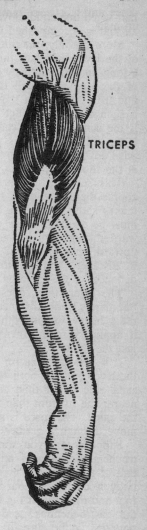

TRICEPS

that when one child moves down the other moves up, and all that is accomplished is amusement.

More usefully, the fulcrum can be placed near one end of the lever, which is thus divided into a short shaft and a long one. A relatively small push downward at the end of the long shaft will then lift a relatively large weight upward at the end of the short shaft. Actually, the balance is such that the product of the force and the shaft length is the same at either end. If, for example, the long shaft is ten times as long as the short one, then a one-pound push downward at the long end will lift a ten-pound weight at the short end.

This seems a cheap way to lift ten pounds, but of course there is a price for everything. The payment one makes shows up in the distance through which the force must be exerted. If the one-pound push is carried downward through ten inches, the ten-pound weight is lifted upward only one inch. What is gained in force is lost in distance, and the total work (which is force times distance) is the same on both sides.

The lever, then, does not increase the work capacity but merely trades distance for force. When it is necessary to lift a heavy weight with a force inadequate for the task if applied directly, then a lever of the sort described is just the thing.

It is also possible for the fulcrum to be at one end of the lever and for both the force and weight to be on the same side rather than on opposite sides. If the weight is nearer the fulcrum than the force is, the situation remains the same. The force is smaller than the weight but must move through a greater distance.

Suppose that you want to stand on tiptoe. The gastrocnemius contracts and pulls your heel upward. The ball of the foot acts as a fulcrum, and the weight of the body is centered about halfway between the ball of the foot and the heel. If the weight of the body is 150 pounds, the gastrocnemius need only pull with a force of 75 pounds. To be sure, it has to lift the heel three inches to raise the body one and a half, but that might seem worth it to save 75 pounds of pull.

And yet it is not always desirable to multiply distance to save force. It is easy to imagine a "reverse-lever" which deliberately multiplies force. Such a lever would have a weight at the end of the long shaft with the force at the end of a short shaft. If the long shaft were ten times the length of the short one, we would have to push downward with a force of ten pounds in order to lift a weight of one pound.

This might seem a ridiculous arrangement, but we gain something too. The one-pound weight at the end of the long shaft moves up ten inches for every inch we push the end of the short shaft downward. Suppose that we tie a heavy weight to the end of the short shaft, a weight almost large enough to balance the small weight at the end of the long shaft. Now we add our own weight to the short shaft, so that the weight at the end of the long shaft lifts through a long arc, at a faster rate than we could move it by the direct application of force. We expend force to gain distance and we have a catapult. The same is true if the weight and force are on the same side of the fulcrum but with the weight farther from the fulcrum than the force is.

Consider the biceps again. When it contracts, it pulls up your lower arm. The bones of the lower arm form a lever with the fulcrum at the elbow. The biceps is attached to the radius about one quarter of the way from the elbow to the palm, so that the force to lift the arm is being applied perhaps three inches from the elbow-fulcrum. The weight that must be lifted, however, is in the palm about twelve inches from the fulcrum. For every pound placed in your palm which you must lift, the biceps must exert a force of four pounds. However, it need only contract one inch to lift the weight four inches, and by multiplying the force, it lends the hand a catapult-like action that makes it possible to develop considerable speed of motion.

It is by such catapult-like actions that a baseball pitcher can throw a fast ball, and horses can move their legs fast enough to race as they do.

The amount of force that can be exerted by muscle is quite astonishing. When you rise from a squatting position, the muscles straightening the knee must exert about ten pounds of force for every pound of weight lifted. Any man who can lift a 200-pound weight on his back (and this is not too hard for a good-sized man in trim condition), exerts a force of 2000 pounds as he straightens his legs, half a ton on each leg.

SOME INDIVIDUAL MUSCLES

Muscles make up about 40 per cent of the weight of a man and about 30 per cent of the weight of a woman. An average man, in other words, will have nearly 60 pounds of muscle and an average woman will have nearly 35 pounds of muscle. (This disparity in muscle weight is explanation enough for the fact that men are more powerful than women—at least in the brute-strength department.) This heavy weight of muscle is necessary in any creature that engages in rapid motion; but in the course of vertebrate evolution, there has been a profound change in muscle distribution, if not in muscle quantity.

Fish move through water by lateral movements of the body, with the tail slapping against the water from either side. The limbs are small and are used merely as balancing and turning organs rather than for propulsion. As a result, it is the musculature of the trunk that is important, and when we eat fish it is the trunk muscles we consume. (That would give us the chance to notice that those trunk muscles are clearly segmented.)

On land the chief means of propulsion is the push of the limbs against the ground, or, in the case of the birds, against the air. Consequently, the musculature of the limbs becomes important and the muscles of the trunk fade out. When we eat meat taken from birds and mammals, it is the limb musculature that we mainly consume, and this is not segmented.

It would be quite tedious to try to list all the muscles in the human body; there are some 650 of them (almost all in pairs), with complicated interrelationships. Nevertheless, it would be helpful to mention a few of the more noticeable muscles.

Starting at the head, there is the *masseter* (ma-see′tur; "chewer" G), which has its origin in the cheekbone and its insertion in the angle of the lower jaw. As the name implies, it is a muscle used in the chewing of food. You can feel it bunch just outside the teeth when you clench your jaw.

Among the many muscles that govern the motions of the head are the *trapezius* (tra-pee′zee-us; "four-sided" G) and the *sternocleidomastoid* (ster′noh-kly′doh-mas′toid; "sternum-clavicle-breast" L, so called because of the bones to which it is attached). The former runs down the back of the neck and pulls the head back, while the latter runs down the side of the neck and pulls the head to the side. Underneath the sternocleidomastoid is the *splenius* (splee′-nee-us; "bandage" G, because of its appearance), which turns the head in a *no* gesture.

The muscle of the upper arm just under the outermost edge of the shoulder is the *deltoid* ("delta-shaped" G, which means "triangular" because the Greek capital-letter delta has the shape of a triangle). It has its origin in the clavicle and scapula and its insertion in the humerus. It is used to raise the arm away from the body. You can feel it harden when you do so.

Opposing the action of the deltoid is the *pectoral* ("breast" G), which underlies the skin of either breast. It has its origin along the clavicle and sternum, and its insertion is also in the humerus. It draws the arm in toward the body, and if you perform this action you will (if you are male) feel the muscle beneath the skin of the breast harden.

Running down the rib cage is a whole series of *inter-costal muscles* ("between the ribs" L) stretching, as the name implies, from one rib to its neighbor. Their function is to expand and contract the rib cage in the course of breathing, and of all the muscles they most clearly show the segmentation common to all vertebrates.

Below the ribs is the abdomen, the largest area of the body to be unprotected by bone. In the average tetrapod this is not serious, for the abdomen is underneath and is the least exposed portion of the body; but in standing upright man has moved the tetrapod's "soft underbelly" into a position of vulnerability. It is vulnerable not only to enemy action but also in the sense that his internal design has worsened.

In the ordinary tetrapod posture the muscles of the abdomen acted as a floor for the intestines and other organs of the interior, and for this purpose it was well designed, with some hundreds of millions of years of evolutionary development behind it. When man rose on his hind legs, the floor became a wall (it is frequently called the "abdominal wall") and for that it is not so well designed. Unless the muscles are kept firm by constant exercise they will bulge flabbily outward and give rise to the unsightly (but very common) potbelly.

Among the muscles of the abdominal wall is the *rectus abdominis* (rek′tus ab-dom′i-nis; "upright of the abdomen" L), which runs from the pubis up to the middle ribs on either side of the midplane of the body. Between these two vertical muscles, a fibrous band runs the whole length of the abdomen, crossing the navel and marking the midplane of the body. It is the *linea alba* (lin′ee-uh al′buh; "white line" L). In lean and muscular men the linea alba can be seen as a shallow, vertical furrow, with the rectus abdominis bulging slightly on either side.

The *transversus abdominis* ("crosswise of the abdomen") runs under the rectus abdominis and at right angles to it, extending from the linea alba around either side of the body. The *obliquus externus abdominis* (ob-ly′kwus eks-ter′nus; "outer oblique of the abdomen") also lines the sides and flanks.

These muscles and others by no means make up an impervious wall. There is constant danger, made worse by man's upright posture and the consequent poor distribution of his internal organs, that a portion of those organs may come to protrude through some weakened section of the wall. The section may be weak congenitally or it may have weakened over the years. Some abnormal strain (lifting a heavy weight, or perhaps a convulsive cough) may finally bring on this situation, which is called a *rupture* (that is, of the integrity of the abdominal wall) or *hernia* ("rupture" L). Hernias usually involve protrusions of sections of intestine, and, although there are a number of possible locations, about 85 per cent are in the groin. These are the *inguinal hernias* (ing′gwin-ul; "groin" L).

At the hip are several large muscles. Among them are the *gluteus medius* (gloo′tee-us mee′dee-us; "middle of the buttock" L) and *gluteus maximus* ("largest of the buttock" L). Both muscles have their origin in the ilium and their point of insertion in the femur. The gluteus medius may be felt on the sides, just under the upper curve of the hipbone, and the gluteus maximus makes up the muscular bulk of the buttocks themselves. You can feel it harden when you pinch your buttocks together.

The gluteus medius has the ability to draw the thigh away from the midplane of the body. The gluteus maximus pulls the thigh into a straight line with the trunk. In other words, when you are sitting down and contract the gluteus maximus (upon which you are sitting), you stand up.

I have mentioned the gastrocnemius as one large muscle of the leg. Another, and even larger one is the *rectus femoris* (rek'tus fem'uh-ris; "upright of the thigh" L), a vertical muscle running from the ilium to the patella and tibia, down the front of the entire length of the thigh. It straightens the leg at the knee when it contracts.

The limbs are essentially solid objects—layers of muscle built about central shafts of bone. The torso, on the contrary, is differently planned. The bone structure is not central but is at the edges. The vertebral column runs along the dorsal edge, the ribs, curve about the sides, and the sternum is at the ventral edge. The pectoral girdle bounds the superior edge and the pelvic girdle the inferior. The muscles attached to those bones also confine themselves to the edges of the torso, filling the space between the bones and producing the abdominal wall as a further boundary at the ventral edge.

Within this muscle-and-bone enclosure is a space in which are to be found the internal organs of the body. In mammals, and in mammals only, the internal space is divided into two parts by a thin partition composed of muscle and tendon and called the *diaphragm* ("partition off" G). It is attached to the sternum in front, to the lower ribs along the sides, and to the vertebral column behind. It bellies upward in its middle stretch so that it divides the body space into a smaller superior cavity and a larger inferior one. The superior cavity is the *thorax* ("chest" G), the inferior one is the *abdomen*. As we shall see, the chief contents of the thoracic cavity are the lungs and heart; the chief contents of the abdominal cavity are the intestines, kidneys, and genital organs.

The diaphragm is not a hermetically sealed partition. A number of blood vessels, nerves, and even a portion of the digestive tract must cross it. It can also give way abnormally, so that abdominal organs may protrude through a weakened section to form a "diaphragmatic hernia."

Not all striated muscles, although called skeletal muscles, are attached to the skeleton. Some are attached to the fascia under the skin. The horse has many such muscles over its body, as we can see when we watch it shake the skin here and there in order to drive away insects. We have lost the capacity for this, but we still retain the use of many muscles under the skin of the face—to an extent more marked than in other animals. It is these facial muscles that make it possible for us to smile, frown, purse our lips, wrinkle our nose, and give our face all the mobile expressivity it has.

We even have small muscles originally intended to move the ear. In creatures such as dogs and horses, these are most useful in directing the outer trumpetlike portion of the ear in the direction of some sound. Our ears are no longer trumpets, and most men cannot use the ear muscles. There are individuals, however, who retain enough of their use to be able to wiggle their ears, an accomplishment that inspires unfailing admiration among those of us who can't.

Structures such as the muscles of the ear or, for that matter, the bones of the coccyx, which represent organs useful to our long-dead ancestors (with trumpet-shaped ears and long tails) but useless to us, are *vestigial organs* (ves-tij'ee-ul; "footprint" L). Like footprints, they indicate that something must once have passed that way.

5

OUR LUNGS

THE ENTRANCE OF OXYGEN

Muscles contraction, and almost all other life processes as well, are energy-consuming. The source of the energy lies in the chemical reactions that go on within the cells, and of these the most important, from the standpoint of energy, are those involving oxygen.

To be sure, current theories as to the origin of life strongly suggest that, to begin with, there was little or no free oxygen available for living organisms on this planet. However, the development of green plants introduced the process of photosynthesis which uses the energy of solar radiation to break up water into hydrogen and oxygen. The hydrogen is used to convert carbon dioxide first into carbohydrates and then into all the other organic components of living tissue. The oxygen is liberated into the atmosphere and, after the green plants multiplied and covered the face of the earth, the atmosphere slowly filled with oxygen.

For at least a billion years, then, the earth's atmosphere has contained a considerable proportion of free oxygen (it is 21 per cent oxygen now). Cells have drawn on it freely during all that time, combining it with foodstuffs to produce energy, while green plants have continued to use solar energy to restore oxygen to the air. The result is a neat balance, which, we trust, will continue on into the indefinite future.

Of course, we think of oxygen mainly as a component of the atmosphere, but that is partly because we ourselves are land creatures living at the bottom of the air ocean and directly dependent on its oxygen content. From the standpoint of breathing, we think of water merely as something to drown in. Yet during most of the span of life upon this planet, the land surfaces were barren and even today not more than 15 per cent of the mass of living organisms dwells on land. All living organisms, not many hundreds of millions of years ago, and most living organisms even now, live in the sea and make no direct use of the oxygen of the atmosphere.

But the creatures of the sea are as dependent upon oxygen as we are. The fact that they live immersed in water means that they obtain oxygen out of the natural waters of the earth by methods for which they are designed and we are not.

Oxygen will dissolve in water. A liter of pure cold water will hold nearly 5 milliliters* of oxygen. Ocean water, which is not pure water, but contains 3½ per cent dissolved solids, can do somewhat better and will hold 9 milliliters of oxygen per liter (0.8 per cent by volume); which through all the vast ocean comes to 10,000,000,000,000,000,000 liters of oxygen. Upon this dissolved oxygen the life of the ocean depends. In water from which the dissolved oxygen has been removed, a fish will drown as easily and as quickly as a man.

The first problem that faces any organism, as far as oxygen is concerned, is getting it out of the environment and into the cell. The cell membrane is semipermeable; that is, it will act to let some substances through and others not; it will even allow some substances through in one direction but not in the other. It is freely permeable in

* A liter is equal to about 1.05 quarts. A milliliter is a thousandth of a liter. These are units of the metric system, used by all civilized nations except Great Britain, the United States, Canada, Australia, New Zealand, and South Africa. Even in these "English-speaking nations" scientists use the metric system in their work. If you are not familiar with the metric system, and want to be, you might try my book *The Realm of Measure* (1960).

either direction, however, to most very small molecules which can be pictured as having no difficulty in slipping through the submicroscopic pores of the membrane. One of the small molecules with this privilege is the oxygen molecule; it *diffuses* ("pour out" L) across the membrane freely. But it diffuses freely in either direction, so it may seem that we have gained nothing. Surely for every oxygen molecule that pops into the cell one from the cell's interior pops out. This might well be the case if the oxygen remained oxygen within the cell.

However, any oxygen that diffuses into the cell is at once combined with the substances within the cell. The oxygen becomes part of molecules incapable of passing through the membrane and is thus trapped. None will diffuse outward. The result is that oxygen travels in one direction only: from the outer environment into the cell.

In general, whenever a substance moves from position A to position B and vice versa, the overall drift is from the point of high concentration to that of low. The difference in concentrations is the *concentration gradient,* and the higher the gradient, the more rapid the overall drift. In this particular case, oxygen travels from the environment where, in the case of the sea, it makes up 0.8 per cent of the volume, into the interior of the cell, where its concentration as free molecular oxygen is virtually zero.

All this is very well, of course, for organisms that consist of a single cell or of a relatively small number of cells, for then the membrane of each cell has environment on one side and protoplasm on the other and diffusion can be depended upon to maintain an adequate oxygen inflow. When we consider fairly large organisms new problems arise. The larger an organism, the larger the proportion of cells located well inside its structure and separated from direct contact with the environment by layers of other cells. The danger of oxygen starvation becomes more serious.

To put it another way, I cite what is called the "square-cube law": if an organism increases in dimensions but retains its shape, its surface will increase as the square of its length, while its volume increases as the cube of its length. To show what that means as simply as possible, let's suppose that a creature one centimeter long has a

surface of one square centimeter and a volume of one
cubic centimeter. A similar creature two centimeters long
would have a surface of 2 times 2 or four square centimeters, but it would have a volume of 2 times 2 times 2
or eight cubic centimeters. We can set up a small table
that will make it even plainer:

Length:	1	2	3	4	5	6	7
Area:	1	4	9	16	25	36	49
Volume:	1	8	27	64	125	216	343

The rate at which oxygen will diffuse into the cell depends upon the amount of surface exposed to its passage.
But the number of cells that the oxygen must supply
depends on the volume of the organism. If a square
centimeter of surface can just barely supply a cubic
centimeter of volume with the needful oxygen, then 49
square centimeters of surface will just barely supply 49
cubic centimeters of volume. If 49 square centimeters of
surface are required to supply the necessary oxygen for
343 cubic centimeters of surface, the creature depending
upon the fulfillment of such a requirement will die.

One solution is for an organism to change shape; to
become longer and flatter, so that more surface is exposed per unit volume. After a certain point is reached,
however, this produces new problems of its own, for the
long thin creature becomes ungainly.

A better and more efficient solution is to specialize at
least a portion of the body for the task of oxygen absorption. Oxygen would be absorbed at a greater rate and this
in turn would make it possible for a given surface area
to support a larger volume. The remainder of the creature's
external surface can then divorce itself altogether from the
task of gathering oxygen and can be made impermeable;
it can be coated with horny scales, bony armor, stony
shells.

To maintain a high rate of absorption of oxygen
through the specialized area, it is also necessary to drive
a current of water past it. Where water is stagnant, the
concentration of oxygen in the water layers near the
absorption area decreases as oxygen passes from those
layers with the cell. This lowers the concentration gradient

and the inflow of oxygen slows. If, however, the water layers next to the absorption area constantly change, the concentration gradient remains high at all times.

Thus the chordates made use of a scheme in which water was drawn in through the mouth and out through slits behind the head. On the way, the current of oxygen-rich water passed across membranes that presented a great deal of thin surface through which oxygen could be absorbed with special ease. These membranes are gills and the slits through which the water emerges are the gill slits. Between the gill slits are the skeletal supports called the gill bars. In sharks, the gill slits are separate and can be clearly seen as vertical clefts just behind the head on either side. In bony fish, there is a *gill cover* over the slits, with an opening behind.

Early in the game an auxiliary means of absorbing oxygen came into play. A buoyancy device is useful to any creature living under the surface of the sea. If a fish is heavier than water, he tends to sink and must work continuously and frantically to keep from doing so. If it were lighter than water, he would tend to rise and would have to work just as continuously and frantically to keep from doing that. It would be most useful if there were some way in which it could adjust its own density so that it could sink, rise or stay put with a minimum of muscular effort.

One answer lay in the development of an internal gas-filled *air bladder* or *swim bladder*. By increasing the gas volume within the bladder, the fish's overall density is decreased; while by decreasing the gas volume, its overall density is increased. The swim bladder opens into the throat region, so the simplest way of adjusting the gas supply in it is for a fish to stick its mouth out of water and swallow some air or discharge some.

But this raises an interesting possibility. The swim bladder is lined with a moist membrane and some of the oxygen in the gulped air will dissolve in the moisture. Such dissolved oxygen will unavoidably diffuse into the cells with which it is in contact and you have what can be called a "lung." This can be tremendously useful. If a fish is living in a body of water which for one reason or another is brackish and low in dissolved oxygen, any additional oxygen that the creature can gain by gulping air and absorbing it through the swim bladder is so much manna. In fact, there is good reason to believe that the bony fish developed in fresh water first, which proved often brackish, and that the swim bladder first became useful as a lung, and only secondarily came to serve as a buoyancy control.

Later fish, which migrated to the oxygen-rich oceans, converted the primitive lung into a pure swim bladder and used it for no other purpose. At least most of them did. Nevertheless, some fish which continued to live in brackish water retained and even refined the lung. There are several species of "lung-fish" living today in Africa, Australia, and South America which can live in foul, muddy water, and even survive for considerable periods in dried mud, by switching from gills to lungs.

About 300,000,000 years ago certain varieties of lung-breathing fish developed into amphibia and during adult life at least abandoned gills altogether. The lungs of the amphibia were rather primitive devices, in comparison with those that developed later among their more advanced descendants. This can be seen plainly in the case of modern amphibia: the adult frog, although using his lung, still absorbs much oxygen directly through the skin—which represents, actually, a step backward.

THE NOSE AND THROAT

Above the mouth in fish is a pair of pits lined with cells equipped to test the chemical content of the water in which it is swimming. (We also possess the same type of cells and we call the sensations to which they give rise "smell.") Among some of the higher fish the nostrils extend backward and open into the rear of the mouth. In this way, food placed in the mouth can be both tasted and smelled, a sense combination more effective than either alone. This situation has persisted in all tetrapods, including ourselves, and what we think of as taste is in fact almost all smell. This proves itself whenever a cold blocks our air passages to the point where the sense of smell is lost. The sense of taste shrinks to almost nothing and meals become a woeful affair.

Once the connection is made between the pits and the mouth, it becomes possible to breathe with the mouth closed. Whereas fish must open and close their mouths constantly to drive water past the gills, a frog breathes with its wide lips closed.

The frog uses the bottom of its mouth as a pump. That area bulges downward and sucks air into its lungs via what were once mere smelling pits; and it squeezes upward to force the air out again. Reptiles and mammals use a more efficient pumping mechanism because they possess ribs, which modern amphibians do not. When the intercostal muscles lift and expand the rib cage, a partial vacuum is created within the chest and air enters from the outside. This is *inspiration* ("breathe in" L). Another set of intercostal muscles then contracts the rib cage, forcing air in an *expiration* ("breathe out" L). Both processes, repeated alternately over and over again, together are *respiration* ("breathe again" L).

In mammals, still another refinement is added in the form of the diaphragm. When the rib cage lifts in mammals, the dome-shaped diaphragm flattens out, further increasing the volume of the chest and hastening the influx of air. When the rib cage presses downward, the diaphragm bulges upward, helping to squeeze out the air.

As in the case of the frog, we can breathe with our mouth closed, the air entering the lungs through that prominent feature in the middle of the face, the *nose* (AS). It enters through twin openings in the nose, each of these being a *nostril* (AS). The nostril actually is an opening drilled into the nose, so to speak, a "nose-drill"— and that is the old meaning of the word. The Latin term for the nostrils is *nares* (nay′reez).

If the nose were nothing more than a mere air vent, there would seem no need for elaborating its structure. Thus, in whales, where it is no more than an air vent, it consists of a single opening (in some species, a double one) set flush with the top of the head, where it can be used for rapid emptying and refilling of the lungs. As far as breathing is concerned, a whale has no time for frills.

Speed is of the essence, and his nasal vent, or vents, is called, dramatically, a "blowhole."

On the other hand, the patterns of life are versatile enough to turn any organ to subsidiary and sometimes surprising uses. A nose can thicken and swell, as in pigs, to become a rooting device. Or it can take on fleshy outgrowths, as in some moles, in order that it may form a delicate organ of touch to substitute for eyes in the dark underground. The nose may even vastly lengthen to become an organ of manipulation, as in the elephant, and become second in versatility and delicacy only to the primate hand.

In man, a compromise has been struck. The nose is still primarily an air vent and has no exotic uses. It nevertheless has a more complex structure than does the whale's simple blowhole; with us breathing is not the emergency it is with the whale. Our lungs can be filled in more leisurely fashion and our nasal passages are lengthened, narrowed, and made more complex in order that air might not only be allowed entrance but that it might be conditioned—moistened and warmed—on its way.

As a result of this lengthening of the passages, the nose forms a definite projection in the midface region. (The nose is one of the more variable of human features, and it contributes a great deal to fixing the general impression a face makes upon us.) Projecting as it does, the nose is in an exposed position, where it can be easily battered and broken by the buffets of the outer world. The septum dividing the nostrils (which ideally marks out an equal division) can be deformed so that a "deviated septum" will result. One or the other of the nasal cavities is narrowed and this can make breathing a bit more difficult.

Our own system of air-conditioning begins at the very entrance to the nostrils. The skin within the opening of the nostrils is hairy, as we can easily see for ourselves, and these hairs serve to strain out any coarse particles, small bugs or other impurities that may be swept in with the air flow. In adult males the hair on the upper lip might serve as an additional safeguard, but this cannot be very important, for women and children (to say nothing of clean-shaven men) lack the mustache and yet do not seem to suffer as a result.

Deeper within the nose subtler devices are used. The air passage takes on a horizontal direction and runs beneath the floor of the cranium in the direction of the throat. Along this horizontal portion of the passage are three bony projections, roughly horizontal and parallel to each other. They are rather intricate in shape and curve sufficiently to remind anatomists of seashells. They are examples of *turbinated bones* ("whirl" L) and are sometimes called simply *turbinates*. More dramatically, they are the *nasal conchae* (kon′kee; "seashell" L). The turbinates divide the air passageways on either side of the nose into three channels, each of which is called a *meatus* (mee-ay′tus; "passage" L). Between the uppermost of the turbinates and the base of the cranium is a recess that contains special cells equipped to give us the sensation of smell. These represent the original olfactory pits of the fish, drawn deep inside the nose.

The air, as it is drawn in through the nasal passages, must make its way through the various meatuses and, in continued contact with the warm, moist walls of the narrow passages, is itself warmed and moistened. Furthermore, because of the curving shape of the turbinates, the air is forced to change direction constantly. Any smaller particles that may have escaped the hairy trap at the nostrils cannot change direction quite as easily as the much lighter air molecules can and at one projection or another the particle is bound to make contact with the nasal lining.

This lining is always stickily moist because it contains *goblet cells*, which secrete a viscous liquid called *mucus* (myoo′kus). Because of this, the lining of the nasal cavities is an example of a *mucous membrane*. Particles making contact with the nasal lining adhere to and are entangled in the mucus. The nasal passages are further moistened by liquid draining from four pairs of hollows in the bones of the face. These are located in the frontal, ethmoid, sphenoid, and maxillary bones and they are called *sinuses* (a Latin term for any hollow with but one opening). The sinuses are lined with tiny cilia that act to swirl the liquid through the narrow openings that connect them with the nasal passages.

In the four-footed position of most mammals, the sinuses are so placed that drainage is downhill. When man tipped himself upright, however, the sinuses tipped with him and took up horizontal or even somewhat uphill positions. Drainage is inefficient and, particularly when the passage to the nasal cavity is blocked during a cold, fluid may accumulate in the sinuses and the pressure then gives rise to excruciating headaches. Anyone who has experienced an attack of sinusitis can be certain that our two-leggedness is not all gravy.

The mucous membrane of the nasal passages also contains ciliated cells. There is a constant beating of the cilia in the direction opposed to the air flow, so that any particles that have escaped all the traps laid for them are beaten back and made to run the gantlet again. The accumulating mucus, with its trapped impurities, can be cleared out of the nasal passages by the force of an explosive expiration of air: the sneeze. This is a reflex

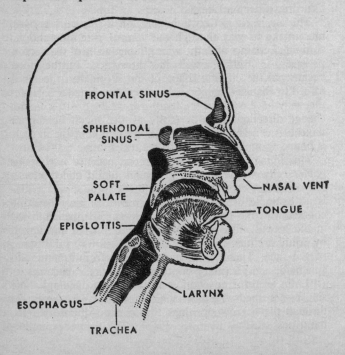

FRONTAL SINUS

SPHENOIDAL SINUS

SOFT PALATE

NASAL VENT

TONGUE

EPIGLOTTIS

ESOPHAGUS

LARYNX

TRACHEA

action induced by any irritation of the nasal membrane and is not under voluntary control, as you know if you have ever tried to stop a sneeze at a time when a sneeze might have been embarrassing. The net result is that the air (if reasonably unpolluted to begin with) enters the lungs in an amazingly clean state. Unfortunately, modern man's tendency to use the air as a disposal dump is too much even for our well-designed nose. Uncounted tons of dust and smoke hang over every large city and the lungs of a city dweller gradually blacken with time.

It is easier to condition the incoming air, by the way, in climates where the atmosphere is warm and moist to begin with. This may be one of the reasons why the Negroes of Africa have developed wide nostrils and a short nose. The inhabitants of Europe have developed narrow nostrils and a long nose, and the longer, narrower passage increases the efficiency of warming and moistening.

Naturally, since the nasal passages are open to the outside world and bear the brunt of its dangers, they are particularly subject to infection. As a result of exposure to wet or to cold or to sudden temperature changes or to sneeze from an already infected person, the virus of the common cold or of influenza can begin to multiply in the nasal passages. The mucous membrane reacts by intensifying its protective functions to the point where they become more troublesome than helpful. A copious flow of mucus gives us the familiar "runny nose." This flow, together with a swelling of the numerous small blood vessels in the mucous membrane, blocks the passages and make breathing through the nose difficult or impossible. Add to that the constant activation of the sneeze reflex, and the misery is complete.

The same set of unpleasant reactions may be in response not to a virus but to a foreign protein which in itself is harmless but to which the body may have developed a sensitivity. During the late summer and fall, for instance, the pollen of many plants floats in the atmosphere. Most of us are unaffected by this. The pollen particles are filtered out as any other particles would be, and there's an end to them. To those with "hay fever," however, contact with the pollen particles throws the respiratory system into a spasm of overprotection. (This is an example of an allergic reaction.) Just as in the case of a cold, there is the runny nose, the congestion of the passages, and the frequent sneezing.

The nasal passage joins the food canal just behind the mouth in the region called the *throat* (AS), or *pharynx* (far'inks; "throat" G). The union of the two passageways at this point makes it possible to breathe with reasonable comfort through the mouth. In fact, when the nasal passages are blocked through a cold or through allergy, it is either breathe through the mouth or die. Nevertheless, the two forms of breathing are not equivalent. The mouth does not possess the adaptations necessary for the proper purification of the air and, except where necessity dictates, it is desirable to breathe through the nose at all times.

Though the two canals join in the throat, they do not remain joined but separate again. Unfortunately, in doing so, they cross. That is, the air passage enters the throat from behind, but below the throat it lies in front of the food canal. This crossing makes it that much easier for either food or air to take a wrong turning. In the case of air, this would not be serious. In the first place, air is not likely to go wrong—it moves naturally in the direction of lower pressure, which means toward the expanding lungs. And then, even if some did find its way into the food canal, it would at the most produce a mild and temporary feeling of discomfort. The matter of food is much more important. Unless there are special precautions, food or drink could be sucked into the air passage by the pull of a low-pressure area just as air could. And with even a small quantity of solid or liquid matter finding its way into the air passages, breathing could be seriously blocked— even to the point of suffocation.

Special precautions against this are, however, indeed taken. The opening of the air passage lies behind and below the tongue and is called the *glottis* ("tongue" G). Just above the glottis is a flap of cartilage attached to the root of the tongue and called the *epiglottis* ("on the tongue" G). In the act of swallowing, when food or water must move through the throat, the glottis automatically moves under the epiglottis, a tight seal being formed. Only one passageway remains open and it is taken by the swallowed material, which moves down toward the stomach and not toward the lungs. You can experience this for yourself, if you begin to swallow and stop at the midpoint; you will find it is impossible to breathe at that stage.

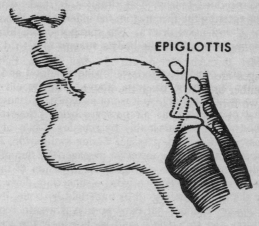

EPIGLOTTIS

The epiglottis and diaphragm combine to produce a sound familiar to all of us. The diaphragm is sometimes subject to periodic spasms of contraction that enlarge the lung cavities and lead to a quick inrush of air. The epiglottis claps down over the glottis to stop the flow, and the air, so suddenly set into motion and so suddenly stopped, makes the sharp noise we call a *hiccup* (AS). The word is spelled to imitate the sound, but by analogy to "cough" it is often written "hiccough," though it is still pronounced "hiccup."

The body does not entirely depend on the perfect functioning of the epiglottis; the matter is too vital. The passageway below the epiglottis is lined with cilia which lash upward to force back and out any tiny impurities that may enter. In addition, any contact of a sizable liquid or solid particle with the glottis sets off an explosive expulsion of air that will blow it out again. This, of course, is a *cough* (AS). When for some reason the epiglottis doesn't get across the glottis in time and we "swallow the wrong way," we go into a strangled fit of coughing that is surely among the unpleasant memories all of us share.

We associate the cough most often with infection, however. The inflammation of the throat that often accompanies a cold causes the mucous membrane of the region to overproduce mucus. The situation is made worse by mucus entering the throat from the inflamed nasal passages above ("post-nasal drip"). The spasmodic coughing that accompanies a cold is the body's attempt to get rid of the mucus.

It is also possible, after swallowing, for food and water to enter the nasal passages above. This would mean motion against gravity, so it is not as likely a misadventure as that of entering the air passages below. Nevertheless, the body guards against this by pressing a flap of tissue against the upper air passage, during swallowing, just as the epiglottis closes the lower air passage. The flap of tissue guarding the upper air passage is an extension of the roof of the mouth, or *palate* (a word of uncertain derivation). Since the roof of the mouth is underlaid by bone, it is the *hard palate*. Behind that hard portion is a soft, rearward extension, the *soft palate,* and at its end is the guarding flap of tissue. You can see that flap if you open your throat wide and look into the mirror. It hangs downward from the top center of the back of the mouth. Because of a fancied resemblance to a grape, it is called the *uvula* (yoo'vyoo-luh; "little grape"). The uvula gives us the curse of the snore. A current of air passing it can cause it to vibrate raspingly. When awake, we automatically keep the passage wide enough to prevent this from happening. At night, some of us do not—as most of us have discovered to our regret.

THE VOICE

Anything that moves is likely to set up the air vibrations we sense as sound. For that reason sound can be an attribute of the inanimate world, as when waves crash against the shore. Or it can be imposed upon living organisms from without, as when the leaves of a tree rustle in the wind. Life itself is distinguished by a variety of noises, from the cricket's strident chirp to the elephant's trumpet.

Noise has its inconvenient side. If it heralds an approach it may prove a warning to an intended prey, so the cat walks on padded toes in order that the sound of its footfall be deadened. More frequently, sound is put to use as psychological warfare, as a code of signals, or, perhaps most important of all, as a mating call. (Considering that an organism must find a member of the opposite sex of its own species from among a swarming myriad of creatures of other species, anything that will increase the chance of discovery and the speed of recognition is useful.)

In mammals a portion of the air passage has been modified for the production of sound. In man this specialized area is found immediately below the glottal opening. It is protected by the *thyroid cartilage* ("shieldlike" G) that nearly encircles it. (The thyroid cartilage is so called because it possesses a downward nick on top of its ventral surface, like a Greek shield of Homeric times. You can feel the nick easily with your fingers if you place them in front of the neck just under the chin.) Immediately beneath the thyroid cartilage is another stff ring, the *cricoid cartilage* (kry'koid; "ringlike" G).

Across the glottis are two folds of tissue called the *vocal cords,* and the space between them is the *rima glottidis* (ry'muh glot'i-dis; "glottal fissure" L). The vocal cords are attached in front to the midpoint of the thyroid cartilage. In back, each is attached to one of a

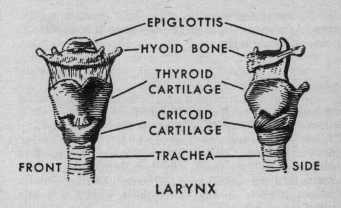

EPIGLOTTIS

HYOID BONE

THYROID
CARTILAGE

CRICOID
CARTILAGE

TRACHEA

FRONT   SIDE

LARYNX

pair of small cartilages called *arytenoid cartilages* (ar'i-tee'noid; "ladle-shaped" G). The cords and cartilages together make up the *voice box,* or to use a Greek term of uncertain derivation, the *larynx.*

Small muscles can rotate the arytenoid cartilages in such a fashion that the vocal cords are moved away from each other, forming a *V* with the apex forward. When this is done, the rima glottidis is wide enough so that air passes up and down without affecting the cords. A rotation in the opposite direction, however, brings the vocal cords close together and parallel. Only a narrow passage is left for air, and as the air current moves past it sets up a vibration in the vocal cords. This is what you are doing when you hum.

VOCAL
CORDS

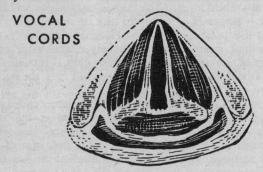

The faster and more forcefully air is expelled past the cords, the louder the hum. Furthermore, it is possible to put the vocal cords under varying degrees of tension. The tenser the cords, the higher the pitch of the hum. We are not aware that we are tightening or relaxing the cords, but that is what we are doing when we sing.

Although with training a singer can learn to do amazing things with his vocal cords in the way of controlled variations of pitch, there is a limit to the extremes he can reach, since he can only tense or relax the cords so far. In general, short cords will produce higher tones than long cords will, and that is why the male voice is generally lower than the female voice. Whereas the rima glottidis in man is up to an inch in length, in women it is only about 0.6 inches.

In children, obviously, the vocal cords are shorter still, accounting for their high-pitched voices and the piercing shrillness of the childish scream. The larynx suddenly increases in size in boys at the beginning of adolescence so that the boyish soprano can become a baritone in rather short order. The change, in fact, may take place faster than the boy can learn to adjust the muscular movements involved in controlling the tension of the cords. The adolescent will find himself, consequently, passing through a period in which his "voice is changing" and he speaks in bursts of baritone and tenor ludicrously interspersed. It is one of the dreadful embarrassments of the early teens. The ultimate size of the adult larynx is not necessarily closely related to the size of the body. A large, strapping athlete may speak in a squeaky tenor and a short, slight man may possess a resonant baritone.

The male larynx is apt to make a prominent pointed bulge in front of the neck, marking the apex of the thyroid cartilage, this being most noticeable in lean men. In women, where the larynx in the first place is smaller and there is a more uniform fatty layer beneath the skin to soften the body outlines, such a bulge is small and not very noticeable.

In swallowing, as the epiglottis moves over the glottis, the glottis (and the larynx that surrounds it) rises in response. When the swallowing is done, the glottis and larynx drop again. The effect, where the bulge of the thyroid cartilage is visible, is a clear (and somewhat humorous) bobbing motion. This motion is visible for the most part in men only, and gives the impression of something swallowed but stuck midway in the neck. This resulted in the legend that when Adam swallowed his bite of the apple in Eden he couldn't quite get it down, and that the mark of this effort is present in all his male descendants in the form of the larynx. This legend, completely unbiblical and unrealistic, is mentioned only because it accounts for the fact that the bulge of the larynx is so commonly referred to as the "Adam's apple."

The nasal passages, the mouth, and the chest all act as a resonating chamber for the sounds produced by the vocal cords. The vocal cords alone would set up a relatively simple and faint sound, and it is to the resonating chambers that we owe the ability to shout. (Some primates, called "howler monkeys," have developed their resonating chambers to the point where their cries can be heard for a mile and more.)

The resonating chamber, moreover, adds a complication of overtones that lend the voice its "quality" or "timbre." Since no two people have nose, mouth, throat, and chest of precisely the same shape, no two people have voices exactly alike in quality. Our ears are designed to detect delicate differences in timbre and it is because of this that we can recognize the voice of a friend or loved one at once, even when it is imperfectly reproduced by the telephone and even when we have not heard it for a long period of time. It is because of this, too, that a mother can quickly recognize the loud cry of her own youngster from the very similar cries of other youngsters in the neighborhood, so that she will react violently and at once to the one, while remaining unmoved by the others.

Man is blessed with an unusually well developed larynx. (In most amphibians and reptiles, in comparison, the larynx is quite crude. In birds, the larynx is present, but the characteristic bird calls are made by another organ, the *syrinx,* lower in the air passages.) It is not man's good larynx that is crucial in the development of human speech but his ability, thanks to the development of his brain and nervous system, to control numerous delicate muscles that alter the condition of the vocal cords and the nature of the resonating chamber.

In particular, the mouth is easily, quickly, and delicately changed in shape by motions of the lips, tongue, and cheeks. The quality of the voice can be changed to produce different sounds. If the air passage remains uninterrupted, the various vowels can be produced. (If you will make various vowel sounds; you will note the manner in which you change the shape of the mouth without blocking the air passage.) If air is forced to make its way through narrow openings, as when you make the sound of *f* or *s,* or is temporarily stopped altogether as when you make the sound of *p* or *k,* consonants are formed. By changing the shape of the mouth continuously and rapidly, a whole spectrum of shifting sounds can be produced in form complicated enough to act as a communication code.

In this manner, abstract thoughts can be expressed with remarkable clarity.

Other animals communicate and some do so not even by sound. (Bees communicate by dancing, and other insects may possess a "mating call" that depends on the sense of smell.) Nevertheless, no land creature but man* can communicate in so complicated a fashion. Even a chimpanzee cannot learn to say more than a few fuzzy, garbled words. Some birds may be able to mimic human sounds and develop a small vocabulary, but they produce the words in a different manner. They tweet them rather than say them, and of course don't understand what they say.

As adults we are not aware of the tremendously complicated muscular movements involved in speech, but this is only because long use has made the whole procedure second nature. Those of us who have reared a child know the years required to learn satisfactory speaking ability.

Infections of the throat and nasal passages will alter the shape of the resonating chambers and change the sound of the voice; roughening it so that we are hoarse. When the membranes of the larynx are themselves affected (*laryngitis*) our speech may be reduced to a whisper. (In whispering, the vocal cords are not involved and use is made of tissue folds, sometimes called "false vocal cords," that lie just above the vocal cords themselves.)

THE BRONCHIAL TREE

Below the larynx is a vertical tube lying just under the ventral surface of the neck. It is about 4½ inches long and is somewhat under an inch in diameter. It is important that this tube be kept open at all times: if the air supply to the lungs is cut off for even a few minutes death will result. For that reason the tube is fitted with C-shaped rings of sturdy cartilage to stiffen the front and sides of the tube. There are 16 to 20 of these rings, separated by fibrous connective tissue. The rings can be tightened by muscles connecting the open ends of the *C*, so that the

* There is growing reason to think that dolphins and related creatures, with brains that are larger and more complicated than our own, may be able to duplicate man's speaking ability.

diameter of the tube can be altered slightly when this is desirable. If you tilt your head back you can feel the tube as you run your fingers down the front of your neck. You will also feel the ridges produced by the alternation of cartilage and fibrous tissue, giving the tube a feeling of roughness. It is therefore called the *trachea* (tray'kee-uh; "rough" G), though a more graphic and less classical word is *windpipe* (AS).

Despite the cartilaginous stiffening, it is still possible to close the windpipe by force. To do this by hand takes considerable effort, however, and the windpipe must be kept closed for several minutes against wild struggling if a victim is to be choked in this fashion. It is not an easy way to commit murder.

A little below the point where the neck meets the trunk, the trachea divides into two branches called *bronchi*

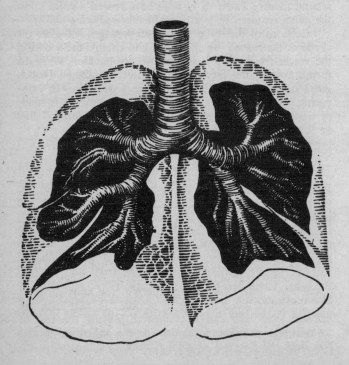

(bron'keye; "windpipe" G). The singular form is *bronchus*. Each bronchus leads to a separate lung, and in doing so divides and subdivides into continually smaller, finer, and more numerous tubes, something in the fashion of the branches of a tree. For this reason, the whole structure is sometimes called the *bronchial tree*. In the larger branches there also is stiffening by rings of cartilage, but as the branches get smaller the cartilage becomes less prominent and finally drops out altogether. When a cold or other infection extends its ravages as far down as the bronchi (*bronchitis*) the coughing spasms that result are particularly agonizing. It is the "cold in the chest" of popular speech as opposed to the "cold in the throat." Bronchitis can very easily become chronic, particularly where the continuous irritation of smoking exists.

The smaller branches of the bronchial tree (*bronchioles*) are themselves a possible source of another sort of unpleasantness. The fine subdivisions of the air passages are lined by circular muscles which through contraction or relaxation can alter the diameter, thus helping to control the air capacity of the lungs. Sometimes, as a result of infection or an allergic reaction to some foreign substance, there is a spasmodic contraction of the small muscles and a swelling of the mucous membrane of the bronchioles. The air passages narrow and the lung capacity is cut down. What follows amounts to a partial strangulation, and although rarely fatal an attack of *asthma* ("to breathe hard" G) can certainly crush out the joy of life.

Surrounding the bronchial tree that stems from each of the main bronchi are the *lungs* (AS). These extend from the collarbone to the diaphragm, one on the right side of the body and one on the left. Taken together, they fill almost all of the thoracic cavity. The two lungs are not quite mirror images of each other. The right lung, which is slightly the larger of the two, is partially divided into three lobes; the left lung is divided only into two. These lobes are, to a certain extent, self-contained. That is, it is possible for one lobe to be suffering a serious disorder and the other lobes to remain normal. In that case the infected lobe itself can be removed, leaving the rest of the lobes to possible normality for the indefinite future. Nor does the removal of a lobe seriously hamper breathing. In many organs there is a built-in "safety factor" where the total work capacity is greater than necessary. In such instances the loss of part of an organ will leave the remainder still able to carry a load sufficient for all reasonable purposes.

The maximum capacity of the lung averages about 6500 cubic centimeters, or about 1.7 gallons. However, what counts in absorbing air is not the volume of the lungs but the area of surface that it presents. Air can only be absorbed at the surface and a quantity of air somewhere in the center of the lungs that happened to be out of touch with the surface could not be absorbed. The fact that it was inside the lungs would make no difference. It might as well be in Africa, for all the absorption it would undergo.

This would indeed be a serious consideration if the lungs were a simple pair of balloon-like organs holding air in a hollow interior lined by smooth walls. If that were the case, the lung surface that would be exposed to the air would amount to something like 2000 square centimeters, or 2 square feet. Lungs are, indeed, more or less simple sacs in lungfish, amphibia, and reptiles, but these are cold-blooded creatures (that is, creatures with body temperatures approximating that of their environment). Cold-blooded creatures, if not necessarily sluggish, can fade into sluggishness when conditions are right.

Birds and mammals, on the contrary, are warm-blooded creatures. Their body temperature is maintained in the neighborhood of 100 degrees Fahrenheit (the normal figure for man is usually given as 98.6 when temperature is measured by mouth) whatever the outside temperature. Not only is this perpetual maintenance of warmth in defiance of the arctic blast energy-consuming, but it keeps our chemical machinery moving at a rapid rate. (All chemical reactions increase in rate with rise in temperature.) To keep the chemical machinery racing, a far larger supply of oxygen is needed than will suffice a cold-blooded creature of similar weight.

For this reason, a lung that is a mere sac would not do for us. Two square feet of surface is not enough. To increase surface area, one must divide and subdivide the interior of the lung, and this is exactly what the bronchi and bronchioles do. The walls of the bronchi and larger bronchioles are far too thick to absorb air but as the bronchioles get smaller and finer, their walls get thinner, until finally, in the smallest bronchioles of all, absorption is possible. These *respiratory bronchioles* lead into several ducts, each of which ends in a cluster of air sacs, each cluster resembling a tiny bunch of grapes. The air sacs end in tiny air cells called *alveoli* (al-vee'oh-leye; "small cavity" L), the singular being *alveolus*.

This whole thin-walled structure, including the respiratory bronchiole, the ducts (usually called *alveolar ducts*), the air sacs, and the alveoli, is called a *lung unit*. It can also be referred to, more dramatically, as the "leaf" of the bronchial tree. The walls of the alveoli are cell membranes, and across those thin membranes oxygen can pass freely. It can pass freely in both directions of course, but the body tissues themselves contain virtually no free oxygen and the air in the alveoli contains a good deal, so that the direction of flow is from the interior of the lungs into the body itself. (Please remember that although the lungs are inside the body the air within the lungs is not. That air is connected by an unobstructed passage to the air outside your nostrils and it is not any more inside your body than the air in the hole of a doughnut is inside the doughnut itself. It is only when the oxygen crosses the alveolar membrane that it enters the body.)

With the volume of the lung broken up completely into lung units, the lung can no longer be considered a sac; it is, rather, a vastly complex sponge. (Because air is retained in this spongy mass even after death, the lungs removed from slaughtered animals are light enough to float on water—the only organs to do so. In butcher-parlance they are therefore referred to as *lights*. The very word "lung" may be distantly derived from the same source as "light." Nevertheless, the human lungs do weigh from 2 to 2½ pound in adults.)

It is estimated that the spongy mass of the lungs contains, altogether, some 600,000,000 alveoli, with oxygen absorption possible over a total surface of at least 600 square feet. Without increasing the total volume of the lungs, then, the surface exposed to absorption has been multiplied 300-fold over the simple sac arrangement. To put it another way, the surface area exposed by the lungs is 25 times the total skin surface of the body.

The lungs, along with the remainder of the respiratory tract, are subject to infectious inflammations. One common lung disease is *lobar pneumonia* ("lung" G). This is often called simply pneumonia, though that term really applies to a whole family of lung inflammations. In an attack of lobar pneumonia, exuded fluid fills the infected section of the lung. Often only a single lobe is involved, whence the name, but when lobes on both lungs are infected, the result is what is popularly known as "double pneumonia."

A second bacterial disease of the lungs is *tuberculosis*. The disease is named for the fact that inflamed nodules called *tubercles* (tyoo'ber-kulz; "little swelling" L) are formed in the infected regions. The alternate name of *consumption* refers to the wasting of tissues that accompanies the disease, which thus gives the impression that the body is being consumed by it. Cancer, which may strike any tissue of the body, is in recent decades finding its way into the lungs with increasing frequency. This rise in lung cancer seems to parallel the large increase in cigarette smoking since World War I and, further, seems to be more common among heavy smokers than among others.

BREATHING

The air within the lungs is not in itself sufficient to maintain life very long. In not very many minutes, its oxygen supply would be gone. It is necessary, then, to renew the air, and this is done by breathing, by forcing old air out and new air in. In the resting adult, the frequency of respiration is 14 to 20 times a minute, depending on the size of the individual. In general, the smaller the person, the faster the normal rate of breathing, so that children breathe more rapidly than adults. (For that matter, a rat breathes 60 times a minute and a canary 108. On the other hand, a horse will breathe only 12 times a minute.)

Even in the short space of time between an inspiration and the following expiration, the composition of the air changes considerably as a result of events in the lungs.

The air you breathe in is about 21 per cent oxygen and 0.03 per cent carbon dioxide. The air you breathe out is 14 per cent oxygen and 5.6 per cent carbon dioxide, on the average. (Not quite all the oxygen lost is replaced by carbon dioxide, as you see. Some is bound up as water and this is usually not counted in working out the composition of air. That is generally calculated on a "dry-weight basis" with the water content subtracted.)

The contents of the lungs are not completely renewed in any one inspiration. In fact, under ordinary conditions of peace and quiet, lung expansion and contraction is gentle indeed. With each quiet inspiration, only about 500 cubic centimeters of air is likely to enter the lungs, and with each quiet expiration 500 cubic centimeters of air leaves the lungs. And even this figure gives a somewhat exaggerated picture of affairs. In an ordinary inhalation the first air to enter the lungs themselves is that which was in the bronchi, throat, and nose—air that had just left the lungs in the previous expiration and had not been pushed out as far as the outside world. Then, after an inspiration is complete, some of the fresh air that entered through the nostrils remains in the air passages, where it is useless, and it is expired again before ever it can get to the lungs. The *dead space* represented by the air passages between the nostrils and lungs amounts to 150 cubic centimeters, so that the fresh air actually entering the lungs with each breath may amount to no more than 350 cubic centimeters. This represents only 1/18 of the lungs' total capacity and is called the *tidal volume*.

The partial replacement of the air within the lungs (*alveolar air*) by the shallow breathing we normally engage in is sufficient for ordinary purposes. We are quite capable of also taking a deep drag of air as well, forcing far more into the lungs than would ordinarily enter. After 500 cubic centimeters of air have been inhaled in a normal quiet breath, 2500 additional cubic centimeters can be sucked in. On the other hand, one can force 700 cubic centimeters of additional air out of the lungs after an ordinary quiet expiration is completed. By forcing all possible air out of the lungs and then drawing in the deepest possible breath, well over 4000 cubic centimeters of new air can be brought into the lungs in one breath. This is the *vital capacity*.*

Even with the utmost straining, the lungs cannot be completely emptied of air. After the last bubble has been forced out, there remain about 1200 cubic centimeters. This is the *residual volume* and is a measure of the necessary inefficiency of the lungs brought on by the fact that those organs are dead ends. (In this respect, birds are more efficient than mammals, for the former possess air sacs in some of the long bones and among the muscles. During inspiration, alveolar air can be pushed out of the lungs altogether and into these air sacs so that the lungs are thoroughly ventilated. This greater efficiency is important to the birds, for flight through air is more energy-consuming than is motion along the ground.)

Breathing is under both voluntary and involuntary control. You can, if you wish, force your lungs to do your bidding. You can take deep breathing exercises or hold you breath completely for several minutes. This voluntary control is important, making it possible to swim under water or to pass through an area filled with noxious vapors. Less obvious is the fact that we could not speak if we did not control the fashion of our breathing. (Try talking without holding your breath and taking quick inspirations between words now and then. It is this breath-holding which forces you to stop to "catch your breath" after you have been talking overrapidly and overlong. And it is one of the necessities of theatrical life that an actor learn to control his breathing in such fashion as to be able to deliver long speeches or sing long arias in a resonant voice without having to stop to catch his breath except at unobstrusive intervals that do not interfere with his performance.)

But it is involuntary control that is chiefly involved in breathing. When we are asleep or when awake but think-

* It is always nice to give figures, since they make a sharp picture of the situation and this usually pleases the reader. However, there is always the danger that the reader will forget the picture isn't really that sharp. There are all sizes and conditions of human beings and no one set of figures holds for all alike. For instance, a small-lunged sedentary female may have a vital capacity of no more than 3000 cubic centimeters, while a trained athelete may have one a high as 6500 cubic centimeters.

ing of other things, breathing follows a regular rhythm at a frequency and intensity designed to bring us the oxygen we need at the moment. This involuntary rate of breathing is governed by the concentration of carbon dioxide in the alveoli. If carbon dioxide concentration rises to a value higher than normal, breathing quickens and deepens. If it falls to a value lower than normal, breathing slows and becomes shallow.

When the voluntary control of breathing brings a person to the point of danger, the involuntary control takes over. Thus, if you deliberately breathe shallowly or hold your breath, you grow quickly uncomfortable. Sooner or later you will give up, and since by underventilating your lungs you have allowed carbon dioxide to accumulate in them, you will—willy-nilly—go through a period of deep and rapid breathing that will lower the carbon dioxide concentration to normal. Even if by great determination you hold your breath until you fall unconscious, the involuntary mechanism will then take over. Children who threaten suicide by breath-holding never succeed.

If, on the other hand, you were to stand on a hilltop and be so overcome by the freshness and sweet country smell of the air that you take to deep breathing, you will eventually flush out almost all the alveolar carbon dioxide. You will quickly become dizzy (*oxygen-drunk*), and while you are sitting down to recover you will virtually suspend breathing for a while.

The breathing rate varies involuntarily for reasons other than your own deliberate interference with your lungs. When the rate of the body's chemical activity increases as a result of physical activity, whether through work, sport, flight, or even through anxiety or nervous tension, more carbon dioxide is produced and poured into the lungs. Breathing automatically becomes deep and rapid; you pant.

If, however, you are existing in enforced quiet and under static conditions (that is, where the stimuli assailing your sense organs are few in number or are monotonous in nature) your breathing will become shallow indeed. This in turn seems to trigger the act of falling asleep. Under conditions where falling asleep would be socially unacceptable, it is desirable to ventilate the lungs thoroughly; the type of forced inspiration that then takes place is the yawn. Unfortunately, society recognizes the yawn as a sign of sleepiness and boredom (which it is) rather than as an attempt to break the shallow-breathing rhythm that leads to sleep (which it also is), so the open yawn is as socially unacceptable as falling asleep. Under conditions where you may neither sleep nor yawn the effort to stay awake may become nothing short of torture.

Lung tissue is elastic and in life is in a stretched state, so its normal tendency is to contract. At birth the baby's lungs are, of course, empty of air. As the baby takes its first breath (traditionally stimulated by a sharp slap on the buttocks while it is held upsidedown by the ankles) the lungs fill and expand, pressing against the rib cage. This contact is not direct. The lungs are surrounded by a membrane called the *pleura* (ploor'uh; "rib" G), which adjoins the lungs, then doubles on itself and adjoins the rib cage. There is, in effect then, a double membrane (with fluid between) along which ribs and lungs can slide against each other with very little friction. Sometimes the pleura grows inflamed through infection (a condition known as *pleurisy*), and breathing can then become agonizingly painful.

When the rib cage lifts and the diaphragm presses down, the lungs are forced to expand even against their own contracting tendency. The alternative would be for the two folds of the pleura to separate with a vacuum between, a situation that is unthinkable. (If you own a small suction disk of the kind that is placed at the end of a child's arrow, trying pressing it against a moistened tile on the bathroom wall and then pulling it directly off by main force, without sliding or peeling. Your difficulty in doing that is an indication of the force that would be required to separate lungs from ribs under ordinary circumstances.)

During expiration the natural elasticity of lung tissue is helpful, and less muscular effort is required to breathe out than to breathe in. If air does find its way between the ribs and the lungs, the elasticity of lung tissue can become a dangerous thing. A wound in the chest may allow air to enter and break the vacuum so that a lung may at once collapse. If the wound is closed, the lung will slowly expand again as the air between lung and ribs is absorbed by the body.

Surgeons sometimes make use of this effect deliberately. When it is necessary for a lung to recuperate from some disease or from surgery under conditions where it is not being continually irritated by the motions of breathing, air can be introduced between ribs and lung so that one lung (not both, of course) can be collapsed for a period of time. Fortunately, the human body can then proceed to get along quite well with only the remaining lung working.

The motions of respiration are fortunately so simple that they can be imposed from without even when the body's own mechanisms fail entirely. This is done by *artificial respiration,* of which there are a number of methods. In some, the victim's lungs are compressed by main force and allowed to relax, over and over again about 10 or 12 times a minute. Or air can be blown into the lungs in mouth-to-mouth fashion. The latter action can be duplicated mechanically by a Pulmotor when it is necessary to continue the action over long periods of time.

Where infantile paralysis has paralyzed the chest muscles, an "iron lung" is used. This is a steel chamber that completely encloses the patient from the neck down. A flexible collar about the neck keeps the chamber airtight. The air pressure within the chamber is raised and lowered rhythmically. At the high point the patient's chest is compressed and breath is forced out; at the low point the patient's chest is pulled into an expansion and breath is forced in. In this way lung action can be maintained as long as the motor works.

6

OUR HEART AND ARTERIES

THE INNER FLUID

What our respiratory system achieves, with all its intricate design, is merely to get the oxygen molecules across the boundary line that separates the environment from the tissues. Where does the oxygen go from there?

When an organism consists of a single cell, there is no problem. Once the oxygen molecules cross the membrane, they are in the cell, where the substances making up the cell's content pounce upon them; and that's that. Even where more than one cell is involved, matters may sometimes be no more complicated. If each cell is equally exposed to the environment (usually to the ocean, sometimes to fresh water), each obtains its oxygen at once by diffusion. Comparatively large organisms can thus exist, provided each cell has its own "ocean front." This means, however, that such an organism can be no more than two-dimensional. Jellyfish and tapeworms are among the longest animal organisms that exist and still depend on diffusion alone for their oxygen. This is at a price. The jellyfish bell consists of a thin outer layer of cells and has nonliving jelly in the interior, while its long tentacles are so thin that no cell is very far from the ocean. As for tapeworms, they

152

are built like tape, as the name implies: long and wide, but flat.

To build up a three-dimensional cellular organization, some cells must be content to be buried in the interior, cut off from the ocean by layer upon layer of other cells. How would the cells in the interior obtain their oxygen? They could not possibly count on diffusion through the hungry cells that lay between their oxygen supply and themselves. The solution was found many hundreds of millions of years ago when some primitive wormlike creatures pinched off a piece of the ocean, in effect, and retained it within the structure of their body. In this way, there is an "internal ocean front," which in time became far more important than the original external one. Eventually, in fact, oxygen absorption from the environment was restricted to a small and specialized portion of the outer body, as I have explained in the previous chapter. The oxygen diffuses through that portion and enters the internal fluid, which is, of course, *blood* (AS).

In relatively small and simple organisms the existence of blood is enough in itself. It can be contained in more or less intricately branched channels so that all cells have a place at or very near the fluid. Oxygen entering the blood will diffuse to all parts of it and, from the blood, into every cell. Diffusion may carry across a relatively long distance but it will not be through layers of oxygen-consuming cells. Every cell will get its fair share.

As an organism becomes larger and more complex, simple diffusion will not do. The length of diffusion is such that parts too distant from the oxygen-absorbing areas might still be starved. It becomes necessary to replace the stagnant pool, so to speak, with a running current that will actively bring oxygen to the cell. It will not be necessary then to rely on the blind, random, and rather slow forces of diffusion. The current is set in motion by a pump in the form of a hollow muscle which, by expanding and contracting, will accept blood and then squirt it powerfully outward. This pump is the *heart* (AS). When blood is forced out of the heart under great pressure, tissues cannot be subjected to the direct blow without damage. The blood leaving the heart must therefore be contained in muscular tubes (*blood vessels*) which absorb the shock and which, by division and subdivision, eventually carry the blood to every part of the body.

In some forms of nonchordate life the blood returns to the heart by filtering its way around the cells. This is slow going, however, and retards the cycle below the point of toleration for large and complex organisms. In chordates (and in some nonchordates) the blood travels by way of blood vessels throughout, both away from the heart and back to it. The blood circulates through a closed system in this way, so that the heart, blood vessels, and blood make up what is called the *circulatory system*. (Actually, the system is not truly closed. There is a type of leakage I shall discuss in the next chapter.)

The structure of the heart is different in the various groups of organisms and, not surprisingly, is more complex in design in the more elaborate creatures. A nonchordate such as the earthworm has a closed circulatory system, and in that system a portion of one of the blood vessels is contractile. A periodic wave of contraction travels down its length, forcing the blood along and driving it forward. This simplest heart, a mere pulsating vessel, is also to be found in a primitive chordate such as amphioxus. Among the vertebrates, however, the pulsating vessel expands into a group of hollow chambers. By enlarging its capacity the pump is capable of squirting blood harder, just as you can blow harder if you take a deep breath first. The enlarged pump naturally develops a muscular wall much thicker and more powerful than that of any blood vessel.

In fish the heart is made up of two main chambers. The anterior chamber is the *atrium* (ay'tree-um; "entrance hall" L, since it serves as one for the more muscular chamber to follow). The atrium contracts and sends the blood into the posterior chamber, which is the *ventricle* ("little belly" L, signifying that it is a small hollow object). The atrium acts as a kind of blood store, collecting it from the incoming blood vessels and then shooting it all at once into the ventricle, which under the stimulus of the sudden muscle-stretching involved in the blood inflow responds by a particularly powerful contraction. When the ventricle contracts, the blood is forced into blood vessels that lead to all the organs of the body. The blood carries oxygen, which is utilized by the cells among which it passes so that by the time the blood returns to the atrium its content of oxygen should be virtually zero. That it is not virtually zero is thanks to the gills. The blood vessels lead to the gills and there pick up oxygen. The oxygen-rich blood from the gills mixes with the oxygen-poor blood from the remaining organs, and the blood contained in most of the vessels is a combination of the two—an oxygen-fairly-rich blood, so to speak.

This was good enough for fish, but the early forms of land life, having developed lungs, began to separate the lung circulation from the rest. The amphibian heart has two atria. The vessels bringing blood from the lungs (oxygen-rich) enter one atrium and those bringing blood from the rest of the body (oxygen-poor) enter the other. The contracting ventricle then alternates at two jobs, sending oxygen-poor blood to the lungs for more oxygen, then oxygen-rich blood to the rest of the body. Mixing of the two types of blood was cut down but not removed entirely.

In reptiles, the ventricle is well on its way to being separated into two portions, and this final step is brought to completion in the birds and mammals. These last, being warm-blooded, used up oxygen at a fearful rate and efficiency simply had to be boosted. In birds and mammals (including ourselves, of course), the heart is four-chambered and consists of two atria and two ventricles. It is really a double pump, bound into one organ and synchronized very carefully. All the blood passes through each pump in turn. Pump number one sends it to the lungs, where it picks up oxygen. The oxygen-rich blood returns to pump number two. It does not mix with oxygen-poor blood at all, but emerges from pump number two in full richness and carries its large oxygen supply to the remainder of the body. In the process, it loses its oxygen, and upon returning to pump number one is sent out to the lungs again. And so the cycle continues.

But now let us turn to the human being specifically.

THE CIRCULATION

The human heart is a cone-shaped organ, about 5 inches long and 3½ inches broad, or roughly the size of the human fist. It weighs about 10 ounces in the adult male, 8 ounces in the female. It lies in the thoracic cavity, just behind the breastbone and between the lungs. Although it is centrally located, its axis of symmetry is not vertical. The conical point is slanted toward the left. This point thrusts out from behind the breastbone and this is the place where the heartbeat can be most easily felt and heard. It is this which gives rise to the mistaken feeling among the general public that the heart is far on the left side.

The heart is essentially a large muscle that is neither skeletal nor visceral in nature but strikes a compromise between the two; so it is put in a class by itself as *cardiac muscle* ("heart" G). Cardiac muscle must have the strength and force of contraction of the skeletal muscles, so its contractile units occur in sufficient quantity to give it a striated appearance (see Chapter 4). However, unlike ordinary striated muscle, cardiac muscle is under completely involuntary control. In that, it resembles visceral muscle. Cardiac muscle differs from skeletal muscle also in that the cells making it up are not separated but interconnect at various places. Such interconnected cells are called a *syncytium* (sin-sish′ee-um; "cells together" G). The heart is composed of two syncytia, one making up the two ventricles and the other the two atria. The existence of these syncytia makes it the more certain that the heart muscle will react as a unit; and in the heart more than in any other organ perfect synchronization of action is important.

The heart, like the lungs, is surrounded by a multiple membrane. This *pericardium* (per-ee-kahr′dee-um; "around the heart" G) is attached to the breastbone in front and to the diaphragm below, and an inner membrane adheres to the heart. There is fluid between the membranes, and when the heart in the course of its beatings moves against the breastbone and diaphragm, the fluid acts as a lubricant to cut down friction.

Let's begin the description of the human circulation at the right atrium. The human atrium is usually called the *auricle* ("little ear" L) because it seems to drape over the top of the heart like a dog's floppy ear, although anatomists prefer the term atrium. Blood enters the right auricle after having passed through all the body, so it contains virtually no oxygen; the heart's first task is to see that this condition is corrected. An efficient mechanism is therefore set into motion.

Between the right auricle and the right ventricle are three little cusps, or pointed flaps, of tissue. These are connected by ligaments to small *papillary muscles* (pap′i-ler-ee; "pimple" L) fixed to the interior of the ventricle. These flaps fold back on those ligaments and offer no bar to blood flowing from the direction of the auricle. Therefore, as blood enters the auricle it pours past those cusps of tissue into the ventricle also. By the time the auricle is full so is the ventricle.

When the right auricle is full its muscular coat contracts and its contents pour past the cusps, adding to what is already there. The distention of the muscular walls of the ventricle as a result of this influx causes it to contract particularly strongly. When it does this, the first surge of blood back toward the auricle pushes the cusps of tissue against the opening, sealing it tightly. Nor can the cusps be pushed through, since the ligaments now hold them tightly against the push of blood. In other words, blood can pass from the auricle to the ventricle past those cusps but not vice versa. The cusps form a one-way valve; this one is referred to as the *tricuspid valve*. There are many other valves here and there in the circulatory system, all working on the same principle and all designed to keep the blood flowing in one direction only. Since the blood from the contracting ventricle cannot push back into the auricle, it must emerge through the only other opening the ventricle possesses. Through that opening it pours into a large blood vessel leading it to the lungs.

The wall of the right ventricle is much more muscular than the wall of the right auricle. Whereas the auricle need only squeeze the blood into a neighboring chamber, the ventricle must force it to the lungs. A harder squeeze is required for the latter.

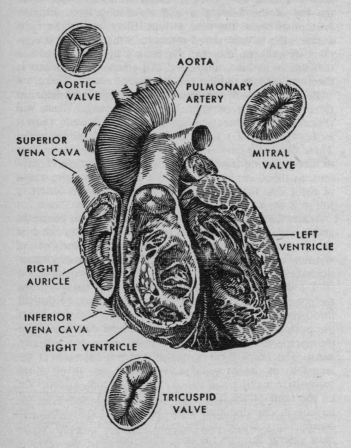

AORTA

PULMONARY
ARTERY

AORTIC
VALVE

SUPERIOR
VENA CAVA

MITRAL
VALVE

LEFT
VENTRICLE

RIGHT
AURICLE

INFERIOR
VENA CAVA

RIGHT VENTRICLE

TRICUSPID
VALVE

SECTION OF HEART
SHOWING CHAMBERS AND VALVES

A blood vessel receiving blood from the ventricle and leading it away from the heart and toward other organs is an *artery* ("air duct" G). This name was given the vessels because they are found empty in dead persons and the early anatomists therefore assumed they carried air. The walls of the arteries are themselves muscular and elastic. As blood surges into one, the muscular wall expands to contain the sudden influx. As in the case of the ventricle itself, expansion is followed by a contraction. There are *semilunar valves* at the artery opening (with the flaps, as the name implies, shaped like half-moons). These will allow blood to enter from the ventricle but will not allow it to return. The contracting artery can thus push the blood in only one direction—farther away from the heart.

The impact of the blood striking the artery wall as it surges out of the ventricle is transmitted along the length of the artery and can be felt at those points where an artery lies just under the skin. The most convenient place for measuring the *pulse* ("to beat" L) is at the wrist, below the palm, and that is where it is traditionally taken. The beat of the pulse is, of course, synchronous with the beat of the heart, and in the days when there was little physicians could do but investigate the most obvious workings of the body, the taking of the pulse was an important diagnostic device. Nowadays much more can be deduced from the heart itself by more subtle methods than the sense of touch alone, so that pulse-taking is no longer all-important. (In motion pictures it seems to have retained its popularity. From a quick feel of the pulse and a close look at the whites of the eyes, the movie doctor seems to make the most amazing and presumably accurate diagnoses.)

The particular artery that accepts the blood from the right ventricle and carries it to the lungs is the *pulmonary artery* ("lung" L). The pulmonary artery quickly divides in two, one branch leading to the right lung, the other to the left. The arteries continue dividing and subdividing, forming smaller and smaller vessels with thinner and thinner walls. The smallest arteries are the *arterioles* (ahr-tee′ree-olez) and these finally divide into *capillaries* (kap′i-ler′eez; "hairlike" L, so named because of their fineness, though actually they are much finer than hairs). This

change is analogous to that of bronchi becoming bronchioles and then alveoli.

The capillaries have walls consisting of single flattened
cells, across which the diffusion of small molecules is easy.
The capillaries arising from the arteries are almost as
numerous as the alveoli arising from the bronchi and, in
fact, along each alveolus there is a section of capillary.
Oxygen molecules crossing the alveolar membrane also
cross the wall of the capillary and find themselves carried
along in the bloodstream. The blood that enters the capillaries of the lung with scarcely any oxygen emerges from
those same capillaries with all the oxygen it can hold.

Gradually the capillaries begin to join into slightly larger
vessels, then still larger ones. Such larger blood vessels
carrying blood back to the heart from the organs are *veins*
(from a Latin word of uncertain origin). The smallest of
these are *venules* (ven'yoolz). By the time the blood is in
the veins, the thrust of the heart pump can be felt no more.
It has been completely absorbed by the friction of the
blood against the innumerable capillary walls. Within the
veins therefore, blood flows much more slowly and more
smoothly than in the arteries. Thus, if a vein is cut, blood
flows out copiously but smoothly. A cut artery gushes
blood in spurts in time to the pumping of the heart. Bleeding from an artery is harder to stop and is far more dangerous than bleeding from a vein.

Since the veins do not have to absorb the shock of heart
action, their walls are thinner than those of arteries and
are not particularly muscular. This loss of pumping action
also means that the blood in the veins is not propelled
forward by the direct action of the heart. The motive force
is instead the pinching action of neighboring muscles as
they contract and thicken in the normal course of their
activity. Many of the larger veins have a series of one-way
valves set along their lengths (particularly veins that must
carry blood toward the heart in opposition to the force of
gravity), and these allow blood to pass only in the direction toward the heart.

The particular vein into which the capillaries and
venules of the lungs finally unite is the *pulmonary* vein.
The pulmonary vein carries the freshly oxygenated blood
to the left auricle, which the blood then enters through a

one-way valve. The pulmonary artery and pulmonary vein, together with all the smaller vessels between, make up the *pulmonary circulation.*

From the left auricle the oxygenated blood passes into the left ventricle through a valve consisting of two, rather than three flaps of tissue. This is called the *mitral valve* (my'trul) because the two flaps, in coming together, seem to resemble the double peak of a bishop's miter, or ceremonial headdress. The two valves, the tricuspid and the mitral, considered together are the *atrioventricular valves,* or *A-V valves,* lying as they do between the atrium (or auricle) and the ventricle.

It is the function of the left ventricle, through its contraction, to send the blood to all portions of the body (except the lungs, which are taken care of by the right

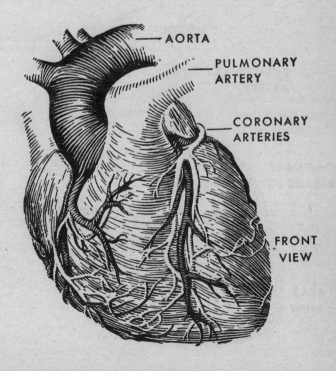

AORTA

PULMONARY
ARTERY

CORONARY
ARTERIES

FRONT
VIEW

ventricle and the pulmonary circulation). The blood must travel farther, after issuing from the left ventricle, and through many more capillaries than is true of the blood leaving the right ventricle, which need travel only as far as the nearby lungs. For that reason and although both ventricles pump out equal quantities of blood with each contraction, the left ventricle must do so with six times the force of the right. It is not surprising, then, that the muscular wall of the left ventricle is twice the thickness of that of the right (another asymmetry of the heart).

The contraction of the left ventricle forces the blood through a one-way valve into the *aorta* (ay-awr'tuh; "to lift up" G, perhaps because the first few inches of the aorta's course lead straight upward). The aorta is the largest

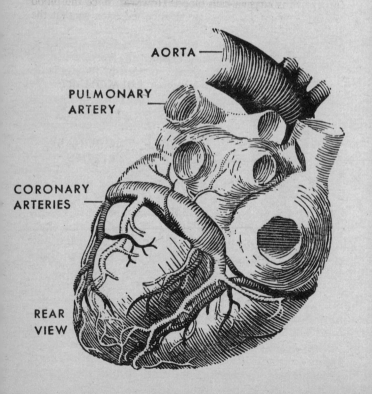

AORTA

PULMONARY
ARTERY

CORONARY
ARTERIES

REAR
VIEW

single artery in the body, with a diameter of a little over an inch at its beginnings. It moves upward at first (the *ascending aorta*), as I have said, but then arches over dorsally (the *arch of the aorta*) and proceeds to run downward (the *descending aorta*) just in front of the vertebral column. In its downward course, the aorta passes through the diaphragm.

From the ascending aorta, at a point just beyond its junction with the left ventricle, two small branches leave it and conduct blood back to the heart. Because these two arteries encircle the heart like a coronet, they are called the *coronary arteries*. It may seem surprising that the heart does not allow itself to be nourished directly by the blood it contains, but it does not. For one thing, only the left half contains oxygen-rich blood. However, once the blood leaves the heart a portion is at once led back so that the heart feeds itself straight from the tap before any other organ or tissue has a chance to get at the blood. Placed in human terms, this may seem selfish of the heart, but if so, it is an enlightened selfishness: its energy requirements are greater than those of any other organ, and upon the heart's smooth and unceasing performance all other organs depend.

From the arch of the aorta, the *brachiocephalic artery* (bray'-kee-oh-see-fal'ik; "arm-head" G) branches off upward. It quickly divides into four chief branches that justify its name. Of these, the two outermost are the *subclavian arteries* (sub-klay'vee-an; "under the clavicle" L, since they travel parallel to that bone to begin with). The subclavian arteries conduct blood to the arms. Between the two subclavian arteries are the two *carotid arteries* (kuhrot'id; "put to sleep" G), which lead blood up either side of the neck. The name arises from the fact that Greek mountebanks would put goats to sleep by pressing that artery and cutting off the flow of blood to the brain.

The descending aorta gives off numerous branches, narrowing as it does so. In the chest region, there are the *bronchial arteries*. These lead to the lungs, but not, as in the case of the pulmonary arteries, for the purpose of picking up oxygen. They already carry oxygen and they serve to supply those portions of the lung, such as the bronchi, with the oxygen they need but which they cannot pick up directly from the air they carry.

A series of arteries, branching off successively lower portions of the aorta, lead to various portions of the digestive tract (with which I shall deal in a later chapter). These include: the *esophageal arteries* (ee'soh-faj'ee-ul) which receive their name from the fact that they lead to the esophagus, the tube connecting throat and stomach; the *celiac arteries* (see'lee-ak; "belly" G), which lead to the stomach and other nearby organs; and the *mesenteric arteries* (mes-en-ter'ik; "middle intestine" G), which lead to the intestines.

Some others include the *intercostal arteries,* which lead to the intercostal muscles, and the *lumbar arteries,* which lead to the lower vertebrae and the muscles of the abdominal wall. The *phrenic arteries* (fren'ik; "diaphragm" G) lead to the diaphragm and the *renal arteries* (ree'nul; "kidney" G) lead to the kidney, as the names imply. (There are other arteries as well, but I won't try to mention them all.)

In the region of the sacrum, what is left of the descending aorta divides into two *common iliac arteries* (il'ee-ak; "groin" G). Each divides into an *external iliac* and an *internal iliac.* The two external iliac arteries supply the legs and the internal iliacs carry blood to the organs of the pelvis.

All these various arteries divide and subdivide into arterioles and, eventually, into capillaries. The capillaries in turn coalesce into venules and then into veins. Usually these return the blood along routes parallel to those of the respective arteries, and have names like those of the arteries. Thus, the blood carried to the kidneys by the renal artery is returned by the *renal vein;* that taken to the hips and legs by the iliac arteries is returned by the *iliac vein;* and so on.

An interesting exception is that the blood carried up through the neck and into the head by the carotid artery is brought back by the *jugular vein* ("throat" L). The jugular vein lies nearer the surface of the throat than does the carotid artery (veins are generally more exposed than arteries, which makes sense since damage to the veins is the less dangerous), and the jugular has therefore become

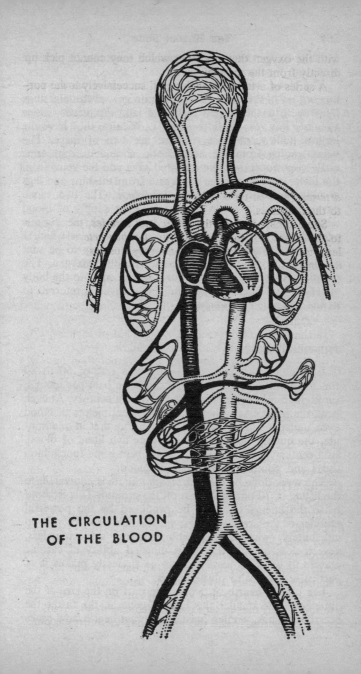

THE CIRCULATION OF THE BLOOD

familiar to the average man as the vein that is severed
when the throat is cut.

The blood, returning by way of the veins from the various organs of the body (except the lungs), eventually finds
its way into the largest veins of all, the *venae cavae*
(vee'nee kay'vee; "hollow veins" L, because they have the
largest hollow, or bore). There are two of these. The
veins leading from the head, neck, shoulders, and arms
coalesce to form the *superior vena cava* and the veins from
the lower portions of the torso and from the hips and legs
combine to form the *inferior vena cava*. The two venae
cavae empty into the right auricle, their supply of blood
depleted of oxygen by the long trip through the various
organs. The blood is now back to the point at which I
began the description of the circulation, ready to be sent
to the lungs once more for a new supply of oxygen. The
circulation leading the blood through the aorta to the body
generally and back through the venae cavae is referred to
as the *systemic circulation*.

THE HEARTBEAT

The human heart beats constantly at a rate of 60 to 80
times a minute, or a trifle faster than one beat per second,
throughout a long life that may last over a century. At each
beat, the heart ejects about 130 cubic centimeters of blood
even under the most restful conditions, so that in one minute, the quietly working heart pumps five liters of blood.
In a century of faithful labor, it will beat some four billion
times and pump 600,000 tons of blood.

The work done by the heart each minute is equivalent to
the lifting of 70 pounds a foot off the ground. This is about
twice the energy that can be produced by the powerful
muscles of the arms and legs; yet the heart can keep it up
indefinitely, whereas the limb muscles, achieving less,
nevertheless quickly tire. This unusual ability of cardiac
muscle to work so hard and yet so tirelessly makes it of
particular interest to physiologists.

The rate of heartbeat depends in part on the size of the
organism. The smaller the size, in general, the faster the
heartbeat. Thus, women have hearts that beat 6 or 8 times

a minute faster than the hearts of men. Children's hearts beat more rapidly still, and rates at birth may be as high as 130 per minute.

This holds true for mammals other than men. The rabbit has a heart that beats at the rate of 200 per minute, and the mouse's tiny heart flutters at 500 beats per minute. Cold-blooded animals that live at a much lower level of internal chemical activity than do birds and mammals get by with a slow beat. The frog, for all its small size, has a heart that beats but 30 times per minute in warm weather, and this beat slows further as the temperature drops. At temperatures near freezing, the rate is down to 6 to 8 per minute.

An animal capable of hibernation shows a remarkable variability of heart rate. A hedgehog has a normal rate of 250 beats per minute, but during the cold weather it survives by going into a cold-blooded suspension of activity during which the heartbeat may drop to 3 per minute. Animals larger than man naturally have slower heartbeats than we do. An ox has a heartbeat of 25 per minute, an elephant has one of 20. The heart rate will also vary in a given creature with his level of activity. During exercise, when the body's requirement for oxygen goes up, the heartbeat is both strengthened and accelerated. The acceleration is brought about also by nervous tension, fear, or joyous expectation and excitement. The "pounding of the heart" under those circumstances is a familiar phenomenon.

Exercise continued over long periods hypertrophies the heart as it would any other muscle. For that reason athletes have a lower rate of heartbeat while resting than sedentary males have. The athletic heart may beat no more than 50 to 60 times a minute. The slower rate is more than made up for by the fact that the heart, enlarged and strengthened through exercise, delivers more blood per beat.

What keeps the heart at its perfect rhythm? The responsibility might be thought to lie with some sort of rhythmic nerve stimulation, but this is not so. Although the heart is indeed outfitted with nerves that can affect its rate, they are not primarily responsible for the beat. This is known from the fact that the heart starts beating in the embryo before it is outfitted with a nerve supply, and it will continue beating in experimental animals even when the nerve supply is cut. The heart muscle will even beat in isolation, provided that it is immersed in an appropriate fluid.

The cells of the heart allow potassium ion to enter but not sodium ion. These ions are electrically charged particles, and by creating a difference in the concentration of such particles within and without the cell an electric potential is set up across the cell membrane. The rise and fall of this electric potential, as ions move across the membrane, sets off a series of contractions (although the details by which this is managed are not yet well understood) and the rhythmicity with which the ions move is reflected in the rhythmicity of the contractions. This means that the working of the heart depends upon the concentration of various ions in the blood, and that these must be controlled within narrow limits. This indeed the body manages; and man himself can imitate this outside the body by using, as I said above, an appropriate fluid.

A heart taken out of the body can be kept alive and beating if it is perfused with a solution containing the proper concentrations of various ions. (By "perfusion" is meant the forcing of liquid through the blood vessels that normally feed an organ.) The first fluid found to be effective for such a purpose was devised by an English physician, Sidney Ringer, and is still known as "Ringer's solution." It is not the intact heart only that beats. A mere portion will beat if properly perfused. It was found, in this way, that different parts of the heart will beat at different rates. That part beating most rapidly, however, forces its rate upon the remainder of an intact heart, for each rise and fall of electric potential moves out along the heart muscle from that most rapidly beating portion and the rest of the heart must follow, having no opportunity to set up potential fluctuations at its own rate. The most rapidly beating part of the heart is therefore referred to as a *pacemaker*.

In the two-chambered heart of the fish, the pacemaker lies in the *sinus venosus* (sy'nus vee-noh'sus; "hollow of the vein" L). This is a widening at the end of the vein leading into the auricle. The beat begins there and progresses down the auricle and ventricle.

The sinus venosus persists in the embryos of birds and

mammals but is gone by time of birth. It fades into the right auricle and its remnants can still be made out there as a bundle of special cells. Because those cells represent the fusion of the sinus venosus and the auricle, they are referred to as the *sinoauricular node,* or, in abbreviated form, the *S-A node.* It is the S-A node that is the pacemaker of the human heart. The wave of electric potential flunctuation that begins at the S-A node spreads out over both auricles (which make up a single fused cell, or syncytium, you may remember), so that both auricles contract simultaneously. There is a momentary pause at the point of division between the auricles and ventricles (the *auriculoventricular node,* or A-V node), where one syncytium ends and another begins. The A-V node, however, soon picks up the wave and sends it along the ventricles, which then also contract simultaneously.

If anything goes wrong with the A-V node, then the beat of the S-A node pacemaker cannot be transmitted to the ventricle, and this is called a *heart block.* This does not mean that the ventricles cease beating. They continue to do so, but at their own natural rate, which is only about 35 times per minute. If the A-V node is in working order, the heart can do better than that even if something goes wrong with the S-A node. The A-V node itself then becomes the pacemaker and keeps the heart beating at a rate of 40 to 50 times a minute.

Sometimes the ventricle contracts prematurely, as a result of some unusual stimulation, perhaps that resulting from some chemical agent in the bloodstream. (Heavy smokers seem particularly subject to this.) When this happens, the prematurely contracted ventricle will not be able to contract when the normal impulse reaches the A-V node an instant later. (After every contraction of the heart, or, for that matter, of any muscle, there is a *refractory period,* during which it will not contract again even when stimulated.) The ventricle must then wait for the next beat. This longer-than-usual wait between beats is the sensation, familiar to some of us, of the heart "skipping a beat." It is not dangerous.

Sometimes, despite the existence of the syncytium, the heart muscle does not contract in proper synchronization. Different fibrils may begin to contract on their own and the result is that the walls of the auricles, for instance, can begin to twitch at a rate of up to 10 times per second. This is *auricular fibrillation* (fy'bril-lay'shun). The A-V node cannot accept beats at this rate (fortunately) and, in effect, generally disregards an auricle gone mad in this fashion. It takes over the ventricular contraction at its own rate which is enough to keep the body going. However, enough of the auricular beats do get across the A-V node now and then to make the overall heartbeat disturbingly irregular. The treatment is usually the administration of digitalis, which acts to depress the conductivity of the A-V node. The ventricle is then less affected by the auricle and the heartbeat becomes slower and more regular. More serious by far is *ventricular fibrillation*. Here, it is the ventricle that begins twitching rapidly. No blood can be pumped and death follows quickly.

Since the heartbeat is so dependent upon the rise and fall of an electric potential, it is not surprising that its rhythm can be upset by an externally induced potential. As a matter of fact, what we call electrocution is usually the result of ventricular fibrillation set up by the electric current passing through the body. As it happens, an alternating current of 60 cycles per second, which is the common type used in the household, is particularly effective in setting up the fibrillation. (The moral is by no means that one ought not to use electricity. Rather, it is that one must be very careful in using electricity.)

A great deal can be told about the workings of the heart by using a device to follow the rise and fall of the electric potential and to measure its progress along the heart muscle. This can be done in animals by placing electrodes directly on the surface of the heart. In humans, one cannot be quite that direct. Fortunately, the tissues conduct electricity and the changes in potential that are associated with the heart action can be detected by connecting appropriate parts of the surface of the body to a galvanometer.

Such a device was first perfected by a Dutch physiologist named Willem Einthoven in 1903. He made use of a very fine fiber of quartz that was silvered in order to allow it to conduct a current. Even tiny potential differences caused noticeable deflections of the fiber and the movements could be photographed. The result is an *electrocardiogram* ("writing of heart electricity" G), which is usually abbreviated ECG. The normal ECG shows five waves, labeled P, Q, R, S, and T. There is first a small rise (P) above the base line, and this represents the motion of the potential wave across the auricle. The passage through the A-V node is represented by Q (a tiny dip below the base), R (a sharp, spiky rise above), and S (another dip below the base, a bit deeper than Q). Finally, the T wave is like the P but is higher and broader and represents the spread of the wave across the ventricle. Changes in the shape and duration of the various waves are useful indications of specific disorders in the heart action.

But before the coming of electrical devices there was the ear, and that can still be used to give important information. The heart is a noisy organ, as we all know. If you place your ear against someone's chest, you. will hear a series of sounds something like: lub—dub——lub—dub ——lub—dub——lub—dub——These sounds arise from the slamming of valves. When the ventricles contract, the tricuspid and mitral valves close and that is the *lub*. When the ventricles relax again, the semilunar valves in the aorta and in the pulmonary artery close, and that is the *dub*. The *lub—dub,* then, mark the beginning and ending of the period of contraction, or *systole* (sis'tuh-lee; "contraction" G). The lapse of time between one *lub—dub* and the next is the period of relaxation, or *diastole* (dy-as'tuh-lee; "dilation" G).

About 1819 a French physician, René T. H. Laënnec, made use of a short wooden tube, one end of which could be placed over the heart and the other to the ear. This made it possible to listen to the heart sounds of women, especially plump ones, without the embarrassment of trying to place the ear directly against the chest. This was the first *stethoscope* ("to view the chest" G, a misnomer since the examination is by ear and not by eye). It has developed into the modern instrument, without which any self-respecting physician would feel quite naked.

The value of the stethoscope (which is today designed to magnify as well as channel the sound) is in its ability to pick up slight variations in sound that indicate abnormalities in the valves. When the damage to valves leaves them unable to close properly (through the scarring that results from a disease such as rheumatic fever, perhaps), there is a backward leakage of blood, or *regurgitation*. When this happens, the clear sound of a valve closing tightly is replaced by a fuzzier sound called a *murmur*. If it is the *lub* sound that is replaced by a murmur, there is regurgitation at one of the A-V valves, usually the mitral, which as part of the left ventricle receives by far the harder blow. If it is the *dub* that is replaced by a murmur, there is a regurgitation at the semilunar valves, usually those of the aorta, which, again, gets the heavier blow.

It is also possible for the valves to be so thickened by scar tissue that they cannot open properly. Even at their widest separation, the opening is abnormally narrow. This is called *stenosis* (stee-noh'sis; "narrowing" G). The blood flows through the opening more rapidly; since the same volume of blood must pass through a narrowed opening in the usual time, its speed must increase if it is all to get through. The blood, foaming over the rough surface of the scarred tissue, also causes a murmur but at a somewhat different point in the sound pattern. The murmur now comes at the point when the valves are open; that is, before the *lub* or *dub* of valve closure.

Valvular disorders are not necessarily fatal, or even very dangerous. They reduce the efficiency of the heart, but not usually by more than the margin of safety. In addition, the heart can compensate by growing larger.

BLOOD PRESSURE

The powerful contraction of the left ventricle causes the blood to surge into the aorta at a speed of 40 centimeters per second (or, to make better sense to the automobile driver, at 0.9 miles an hour). If the aorta were to narrow, the rate of motion would increase, since the same volume of liquid would have to pass through the narrowed diameter in a given unit of time, and the only way that could be done would be to increase the speed of passage. By the same line of reasoning, the speed would decrease if the aorta were to widen.

If you were to examine the aorta as it passes down the

midplane of the body, you would find that it does indeed narrow in diameter, but it is also giving off branches to drain off some of the blood. What counts now is not the width of any one vessel but the sum of the cross-sectional areas of all the different branches. As the aorta divides and subdivides, the individual branches are narrower and narrower but the sum of the areas continually increases. By the time the blood has made its way into the arterioles, the total cross-sectional area of the various vessels through which it is passing is 15 to 30 times that of the aorta and the blood flow is only about 2 centimeters per second.

In the capillaries, which are individually so thin as to be invisible without a microscope, the total cross-sectional area is nevertheless about 750 times that of the aorta, and the blood is creeping along at the rate of but half a millimeter per second. At this slow motion, the blood of the capillaries along the alveoli of the lungs has ample time to pick up oxygen and, in the remaining tissue, has ample time to give off oxygen. Collecting into venules, the total area decreases and the velocity picks up again. The two venae cavae, taken together, have an area four times that of the aorta, so blood re-enters the right auricle at a speed of about 10 centimeters per second.

When blood is forced into the aorta, it exerts a pressure against the walls that is referred to as *blood pressure*. This pressure is measured by a device called a *sphygmomanometer* (sfig′moh-ma-nom′i-ter; "to measure the pressure of the pulse" G), an instrument which, next to the stethoscope, is surely the darling of the general practitioner. The sphygmomanometer consists of a flat rubber bag some 5 inches wide and 8 inches long. This is kept in a cloth bag that can be wrapped snugly about the upper arm, just over the elbow. The interior of the rubber bag is pumped up with air by means of a little rubber bulb fitted with a one-way valve. As the bag is pumped up, the pressure within it increases and that pressure is measured by a small mercury manometer to which the interior of the bag is connected by a second tube.

As the bag is pumped up, the arm is compressed until, at a certain point, the pressure of the bag against the arm equals the blood pressure. At that point, the main artery of the arm is pinched closed and the pulse in the lower arm (where the physician is listening with a stethoscope) ceases.

Now air is allowed to escape from the bag and, as it does so, the level of mercury in the manometer begins to fall and blood begins to make its way through the gradually opening artery. The person measuring the blood pressure can hear the first weak beats and the reading of the manometer at that point is the *systolic pressure*, for those first beats can be heard during systole, when the blood pressure is highest. As the air continues to escape and the mercury level to fall, there comes a characteristic quality of the beat that indicates the *diastolic pressure;* the pressure when the heart is relaxed.

The blood pressure, unlike the rate of heartbeat, is roughly the same in all warm-blooded animals, regardless of size. The systolic pressure is in the range of 110 to 115 millimeters of mercury while the diastolic pressure is about 80 millimeters of mercury. (Atmospheric pressure is 760 millimeters of mercury so that systolic pressure is about 0.15 to 0.20 of an atmosphere, while diastolic pressure is about 0.10 of an atmosphere.)

Blood pressure is not a constant factor in man. For one thing, it varies with age. In a newborn baby, the systolic pressure is no more than 40 millimeters of mercury, though it rises to 80 by the end of the first month. It continues to rise more slowly, reaching 100 millimeters at the start of adolescence and 120 millimeters in the later teens. There is a continued very gradual rise during later adulthood. At the age of 60, a blood pressure of 135 systolic and 90 diastolic would seem quite normal. Exercise and nervous tension also increase the blood pressure, as would seem logical. Where the body generally requires a greater supply of oxygen and the heart, in consequence, beats more rapidly and forcefully, the pressure of the blood on the artery walls will rise. Systolic pressures up to 180 or 200 millimeters of mercury would not be unusual or worrisome as a temporary phenomenon.

The elasticity of the arteries tends to lower the systolic blood pressure, for as they belly out to receive the influx of blood, more room is made for the volume and a smaller push is exerted against the retreating walls. With increasing age, however, arteries lose flexibility, as calcium salts are deposited in their walls, converting them (sometimes) to nearly bone-hard tubes in old age. This is *arteriosclerosis* (ahr-tee′ree-oh-skle-roh′sis; "hardening of the arteries" G). Under these conditions, systolic pressure rises and the slow hardening of the arteries may account for the normal slow rise in systolic pressure in the years after maturity.

Temporary changes in blood pressure can be induced by contraction of the arterioles, the muscular walls of which can close those small vessels altogether. This contractility of the arterioles serves a useful purpose in shifting the distribution of blood to meet the changing needs of the body. Ordinarily, at rest, 25 per cent of the blood is passing through the muscles and another 25 per cent through the kidneys. In addition, 15 per cent is passing through the intestinal regions, with another 10 per cent through the liver. Further, 8 per cent passes through the brain, 4 per cent through the blood vessels feeding the heart, and 13 per cent through lungs and elsewhere.

Under conditions of fright or anger, it is important that the lungs, heart, and muscles be well supplied with blood. The intestinal regions, on the other hand, can temporarily do without; time enough for the slow processes of digestion after the emergency has passed. Through the contraction of appropriate arterioles, the intestines are deprived of some of the blood, which is then distributed through more essential regions.

A more clearly visible display of the variability of blood distribution involves the skin. The skin is well supplied with blood vessels, not merely to supply its cells with needed nourishment, but as a device to carry heat from the interior of the body to the surface, where it may be conducted or radiated into the atmosphere. On warm days, particularly humid ones, or at times when heightened muscular activity produces heat at a greater rate than usual, the vessels in the skin relax. This is *vasodilation* (vas′oh-dy-lay′shun; "vessel dilation" L). There is then room for a greater fraction of the blood in the skin, and loss of heat to the atmosphere is increased. We are visibly flushed as a result on a hot, muggy day or after a strenuous session of work or play. Emotional factors can also produce vasodilation of the skin vessels so that we blush with embarrassment, confusion, shame or, sometimes, pleasure. But on

cold days, when it is necessary to cut down the loss of
heat to the atmosphere, the blood vessels of the skin will
constrict (*vasoconstriction*) and the skin will contain less
than its normal content of blood. We can then turn white
with cold. Emotion can cause the same color change, this
time making us turn pale with fear or shock.

Large veins, particularly of the abdominal region, can
also contract so as to contain less blood and thus make
more of it available for the capillaries of the muscles and
other key organs. Most of all, there is the spleen, a
brownish-red organ lying on the left side of the body just
behind the stomach. It is about the size of the heart but
is not nearly as compact, weighing only 5 or 6 ounces. Its
spongy structure serves as a blood reservoir. It can expand
to hold a liter of blood and, in case of need, contract to
squeeze all but 50 millimeters of its blood supply into the
general circulation.

All these devices can be used to change blood volume or
blood-vessel volume (or both) and thus change the blood
pressure, but under normal conditions a rise in blood pressure is only temporary and is intended only to meet a
temporary need. Sometimes, however, blood pressure rises
and remains high more or less permanently. The systolic
blood pressure may reach 300 millimeters of mercury, the
diastolic 150, both values being roughly twice the normal.
This is *hypertension* ("stretched beyond" G), or, in direct
English, *high blood pressure*. This is dangerous for numerous reasons. It places an unusual strain on the heart and
arteries, contributing to degenerative changes in their structure. The smaller arteries, damaged by the constant high
pressure on the walls, may undergo unusual hardening and
be less able to adjust to the high pressure than ever. They
may even rupture.

The rupture of an artery in the brain is a particularly
serious phenomenon, since, if a sizable portion of the
organ is harmed as a result, paralysis or death follows
quickly. The unfortunate victim is struck down so quickly
and so without warning, in fact, that the condition is
termed a *stroke, apoplexy* (ap'oh-plek'see; "a striking
down" G), or *cerebral hemorrhage*. Naturally, this is more
likely to happen at a time when excitement or overwork
raises the blood pressure beyond even its usual level.

Sometimes hypertension is brought about by a failure of the machinery by which the kidneys control blood pressure (I shall discuss this in a later chapter) in which case physicians speak of *renal hypertension*. Often there is no known cause, and it is then called *essential hypertension*. One of the meanings of "essential" in the medical vocabulary is "without known cause." A synonym for "essential" used in this fashion is "idiopathic" (id′ee-oh-path′ik; "individual suffering" G), that is, suffering about which no general statement can be made.*

Arteries do not degenerate with age, only through hardening as the result of deposition of calcium salts. One change that can occur in middle age, and one equally disastrous, is the deposition of certain fatty compounds on the inner lining of the arteries. The ordinarily smooth lining of the inner walls is roughened by such deposition and takes on an irregular appearance that seemed to some investigators to be like the grains of cooked porridge. The disease is therefore called *atherosclerosis* (ath′ur-roh-skle-roh′sis; "porridge-like hardening" G).

The atherosclerotic artery is dangerous for two reasons. In the first place, the rough coating can damage small bodies in the blood whose function it is to initiate clotting. There is therefore always the chance of clots forming in such an artery. A clot may break up without damage after being formed, or it may be whirled along by the bloodstream until it comes to an artery too small to pass through —in which case it may plug the vessel and stop that portion of the blood flow. This is *thrombosis* (throm-boh′sis; "clot" G). A thrombosis in an arteriole of the brain is as effective as a vessel rupture in bringing on a stroke. Secondly, the atherosclerotic deposition narrows the bore of the artery, sometimes to an alarming extent, and decreases its flexibility as well. For both reasons, pressure goes up within that artery and blood flow is even so decreased.

Particularly vulnerable to these changes are the coronary arteries. This is not because the coronary arteries are unusually weak, but because the heart's needs are abnormally

* The 20th edition of *Stedman's Medical Dictionary*, a very valuable but generally humorless work, unbends far enough to comment as follows upon the term "idiopathic": "a high-flown term to conceal ignorance."

high, so that there is less margin for safety. Whereas most organs make use of but one fourth of the oxygen supply of the blood pushing through them under ordinary conditions, the heart makes use of four fifths of the available supply. Another organ might make do with a diminished blood supply without too much trouble; the heart cannot.

When the constricted coronary artery cannot carry enough blood, there is a sharp pain in the chest. This may also be felt in areas distant from the actual organ affected ("referred pain"), particularly in the left shoulder and arm. The condition is *angina pectoris* (an-jy'nuh pek'tuh-ris; "a choking in the chest" L). Usually an attack of angina follows a bout of work or a period of emotional tension, when the heartbeat is increased and the heart's demand for blood becomes distinctly greater than the constricted coronaries can supply. Drugs such as amyl nitrate or nitroglycerin are sometimes used in these cases, since they act to bring about a general relaxation of arteries, increasing the blood supply to the heart.

Where a clot blocks one of the divisions of the coronary artery, the result is *coronary thrombosis* (the familiar "heart attack"). This can be quickly fatal, but if the blocked artery is small enough it may result only in the death of that portion of the heart immediately fed by that artery. Scar tissue will form in that spot, and after recovery this need not of itself seriously damage a person's life expectancy—except, of course, that the conditions giving rise to one thrombosis remain, and may in all likelihood give rise to another and more serious one.

Vessels may sometimes become abnormally dilated through damage to their walls. A hardened aorta may undergo damage in a particular spot which will then heal under the stretched conditions imposed by the pounding of the high-pressure blood. There is then a permanent weakness in the wall, which bellies out with each beat of the heart. This is an *aneurysm* (an'yoo-riz-um; "wide" G). The danger is that the aorta may simply rupture after one heartbeat too many, with death as a consequence.

Veins, too, may become overdilated. There the damaging factor is not the blood pressure, which is comparatively low in the veins, but the force of gravity. Blood returning to the heart from the legs and hips must move against gravity when a person is standing or sitting. This motion against gravity is accomplished by the normal muscular action which pinches the blood of the vessels toward the heart, thanks to the one-way valves in the veins of the legs. If anything happens to damage the valves or make them ineffective, the return of the blood is seriously hampered. Blood collects in the veins, which distend to four or five times their normal diameters and the result is *varicose veins*. The case is naturally aggravated where a person's way of life involves much standing and little activity.

These various disorders of the circulatory system which I have been mentioning in this chapter are of particular importance nowadays. In recent decades, with many infectious diseases that were once deadly scourges of humanity coming under control, the various circulatory disorders have become the major causes of death in the United States. Nearly a million people a year die of some malfunction of the heart or vessels, and this represents about 55 per cent of all deaths in this nation.

7

OUR BLOOD

The importance of the heart and blood vessels is not so much in themselves as in what they carry, because all their intricate machinery is merely designed to make it certain that each part of the body is properly bathed in a current of blood. The total quantity of blood in a human being is sizable, it being estimated that something like 1/14 of the weight of the body is blood. Men are in this sense bloodier than women; the average man contains about 79 milliliters of blood for each kilogram of body weight, but the corresponding figure for a woman is only 65. A man of average size will therefore contain about 5.5 liters of blood (or about 1½ gallons); an average-sized woman will contain 3.25 liters of blood (or about ⅞ of a gallon).

The most obviously unusual thing about blood is that it is a liquid, whereas other tissues of the body are solid or semisolid. And yet this does not mean that blood is unusually watery. The body as a whole is about 60 per cent water. Considering that life began in the oceans, this is not surprising. On land as in the sea, the chemical reactions in the cells are carried on against a watery background now just as they were when the first living molecule combinations arose in the ocean. If there is to be astonishment, it is that land life has managed to economize on water to the extent of making do with but 60 per cent. Some non-chordate ocean creatures have water contents running up to 99 per cent.

One factor that keeps man's water content as low as it is is that certain relatively inactive tissues can afford to be fairly dry. The fat stores of the body, for instance, are only 20 per cent water, while marrow-free bone is only 25 per cent water. If we consider only the "soft tissues" of the body—the ones in which the chemical business of the body is being actively carried on—the water content runs from 70 to 80 per cent. The liver, for example, is 70 per cent water and the muscles are 75 per cent water.

Blood is 80 per cent water, but this comparatively high value is not the reason for its liquid state, for the kidney, a solid organ, is also 80 per cent water. In fact, the most watery tissue of the body is the gray matter of the brain, which, though certainly not a liquid, is 85 per cent water. The question we should really ask is this: If blood is a liquid, why is its water content lower than that of the gray matter of the brain and almost as low as muscle's? The answer is that although blood may have begun long eons ago as a pinched-off portion of the ocean, its composition is now, as a result of many intricate evolutionary changes, far more complex than the composition of the ocean is, or ever was.

To be sure, there remain important similarities between the blood and the ocean even yet. The blood contains the same ions the ocean does, and in roughly similar proportions. As in the case of the ocean, the most important ions in blood are sodium ion and chloride ion. This gives the blood (and the ocean) its salty taste.

In addition to inorganic ions, however, blood contains organic constituents, complex carbon-containing molecules formed by the body, such as glucose (a kind of sugar), and a large variety of proteins.* What's more, the blood contains objects of cellular size. Some of these are actually true cells. Others are not, being smaller than the average

* In this book it is not possible to go into any great detail concerning the biochemistry of the blood. If you are curious about the matter, you can find a fairly thorough discussion of the subject in my book *The Living River* (1959).

cell and lacking nuclei. Nevertheless, these latter objects remain far larger than any molecule, and are marked off from the truly liquid portion of the blood by membranes. These cells and subcells are lumped together as *formed elements*.

It is these formed elements which give blood its viscosity. If the formed elements swelled somewhat and adhered to each other, as cells do in other tissues, blood would be a soft semisolid as brain, kidney, and muscles are. It is because the formed elements do not adhere to each other but float individually in the blood that blood remains fluid. The formed elements are large enough to settle out easily under centrifugal force. Special graduated tubes are filled with blood (to which a small quantity of chemical is added to prevent clotting) and spun at the rate of 50 revolutions per second. Under such conditions, the formed elements are forced away from the center of rotation to the bottom of the tube and compressed there.

In this way, the blood is divided into two parts: a watery portion called the *blood plasma* and the formed elements. Blood itself, consisting of both plasma and formed elements, is sometimes called *whole blood* to emphasize the fact that all its contents are being referred to. The blood plasma is a straw-colored liquid, 92 per cent water. It is the true fluid of the blood, and in it the formed elements float. The plasma makes up roughly 55 per cent of the volume of whole blood under normal conditions. The formed elements make up the remaining 45 per cent, this percentage being the *hematocrit* (hem'uh-toh-krit'; "to separate blood" G).

The blood has numerous functions, among which one of the most important is that of transporting oxygen. When simple organisms first made use of an internal fluid for the purpose of bathing the inner cells of the organism with dissolved oxygen (and other substances), there remained a serious problem. Water does not really dissolve much oxygen. A liter of ice water will only dissolve 14 milligrams of oxygen out of the air, and this solvent capacity of water actually decreases as the temperature rises. At the temperature of the human blood, a liter of water will dissolve only 7 milligrams of oxygen out of the air. For a simple organism consisting of one cell or a small group of cells, this small solvent capacity of water for oxygen is nevertheless ample, simply because of the huge volume of the ocean. The oxygen supply is virtually unlimited when you think that a cubic mile of ocean water contains up to 60,000 tons of dissolved oxygen and that there are hundreds of millions of cubic miles of water in the oceans.

The situation changes drastically when it is more than a question of individual cells, or even small conglomerations of cells, floating in all those cubic miles, but billions upon billions of cells all depending upon an internal fluid of very limited quantity. If our own blood carried oxygen only by dissolving the gas in its water content, it would at no time carry more than about 30 milligrams of oxygen. This is approximately a 4½-second supply as far as our minimum body needs are concerned, and on such an insignificant margin of safety no complex organism could exist. Some small insects make out with the oxygen dissolved in their water content alone, but certainly we cannot. The mere fact that we can hold our breath for a minute or two without undue trouble is proof enough that our blood carries oxygen through some device more efficient than mere solution.

To get around this problem, we and all other sizable multicellular animals make use of compounds, more or less complex, that are capable of forming loose associations with oxygen molecules. At the gills or lungs, these compounds take up oxygen (and far more oxygen can be held in a given volume of blood in this fashion than through simple solution). At the tissues, the weak hold of the compounds on the oxygen is broken and the oxygen diffuses into the cells. These oxygen-holding compounds are usually colored (although the color has no direct connection with the problem of transporting oxygen) and so they are generally referred to as *respiratory pigments*. These pigments are protein in nature, which means that they consist of large and complex molecules made up of thousands, and sometimes hundreds of thousands, of atoms of carbon, hydrogen, oxygen, and nitrogen. In addition, they almost invariably contain one or more atoms of some metal in each molecule.

The metal involved is most usually iron, but many

crustaceans and mollusks make use of a respiratory pigment containing copper. This copper-containing compound is *hemocyanin* (hee'moh-sy'uh-nin; "blue blood" G) and because the compound is blue in color, the creatures making use of it have blue blood indeed. Those very primitive chordates the tunicates possess a respiratory pigment containing vanadium and certain mollusks possess one containing manganese, but this is very unusual.

The respiratory pigment of the human being, and indeed of all vertebrates, contains iron and is called *hemoglobin* (hee'moh-gloh'bin), for reasons I shall explain later. There are a number of other iron-containing respiratory pigments elsewhere in the animal world but none as efficient as hemoglobin. Nevertheless, the compound is not confined to the vertebrates. As lowly a creature as the earthworm is our cousin in this respect, for it, too, possesses hemoglobin.

The molecule of hemoglobin contains, roughly, 10,000 atoms and has a molecular weight of 67,000. (That is, its molecule weighs 67,000 times as much as a single hydrogen atom, which is the lightest of all atoms.) Most of each molecule of hemoglobin is made up of amino acids, relatively small compounds that go into the structure of all proteins. However, each molecule also contains four groups of atoms not amino acid in nature. These groups contain atoms arranged in a large circle made up of four smaller circles (a very stable arrangement in this particular case) and at the very center is an iron atom.

This iron-containing portion can be isolated from the hemoglobin molecule and is called *heme* (heem), so we can say that there are four heme groups to each hemoglobin molecule. (The Greek word for "blood" is *haima*, which gives rise to numerous terms in connection with blood. The stem derived from *haima* is spelled "hem" in American usage, and "haem" in British usage. I shall follow the American spelling.)

Now in some nonchordate creatures the respiratory pigment is carried in solution in the plasma. This is so in the case of hemocyanin and of some iron-containing pigments. Those creatures possessing hemoglobin, however, always retain the pigment in small containers. This is certainly true of ourselves, and these hemoglobin containers are by far the most numerous of the formed elements I mentioned earlier.

THE ERYTHROCYTE

Hemoglobin in one of its forms is a bright red compound and lends that color to its container, the *erythrocyte* (ee-rith'roh-site; "red cell" G), which thereby receives its name. These particular formed elements are often called, in plain English, *red blood cells,* or simply *red cells.* To be sure, the individual erythrocyte is not red but straw-colored. In quantity, however, the color does deepen to red, and it is this that lends blood as a whole its red color.

There is some question, though, whether the erythrocyte ought to be called a cell, since it contains no nucleus. It is therefore often called a *red blood corpuscle* ("corpuscle" comes from a Latin word meaning, neutrally enough, "little object"). The erythrocyte is smaller than the average cell, having a diameter of 7.2 microns. (A micron is equal to a thousandth of a millimeter, or, if you prefer, 1/25,000 of an inch.) Furthermore, it is disk-shaped, being only 2.2 microns thick. It is narrower in the center, so that it may be described as a biconcave disk. This thinness places the hemoglobin content of the erythrocyte nearer the surface and facilitates the uptake of oxygen.

(When erythrocytes were first studied by microscopes, those early instruments weren't good enough to show the shape precisely. The erythrocytes seemed simply tiny spheres at the limit of vision and were called "globules." Proteins obtained from them were called "globulins" or "globins" in consequence, and it is to that misconception that we owe the word "hemoglobin.")

If the erythrocyte is not a complete cell, it at least begins life as one. It is originally formed in the bone marrow of the skull, ribs, and vertebrae, and, in children, in the marrow at the end of the long bones of the arms and legs as well. The process of erythrocyte formation is *erythropoiesis* (ee-rith'roh-poy-ee'sis; "making of red cells" G). What is later to be an erythrocyte is at first an ordinary, rather large cell, complete with nucleus but containing no hemoglobin. At that early stage it is a *megaloblast* (meg'uh-loh-blast; "large bud" G), since it is the bud, so to speak, out of which the erythrocyte eventually flowers. The megaloblast gains hemoglobin and becomes an *erythroblast* ("red bud" G). Then as it divides and redivides, it shrinks in size and becomes a *normoblast* ("normal bud" G), because it is then the normal size of an erythrocyte. But at that stage it still has a nucleus and is still a true cell.

At the next stage, it loses its nucleus and it then becomes a *reticulocyte* (ree-tik'yoo-loh-site; "network cell" G), because when properly stained its surface can be made to show an intricate network pattern. The reticulocyte is discharged into the bloodstream, and within a few hours it has become a full-fledged erythrocyte. In normal blood, one cell in every two hundred is still in the freshly formed reticulocyte stage. In cases where, for one reason or another, it is desirable to stimulate an increase in the rate of formation of erythrocytes, the earliest sign that the treatment is succeeding is an increase in the proportion of reticulocytes in the blood. This is the *reticulocyte response*.

Erythropoiesis produces an almost incredible quantity of erythrocytes. A single drop of blood contains perhaps 50 cubic millimeters and in each cubic millimeter of male blood there is an average of 5,400,000 erythrocytes. The corresponding figure for female blood is slightly less— 4,800,000.* This means that an average man would possess

* Although the average woman has less blood for her weight than the average man, and although what blood she does have is less rich in erythrocytes, this does not make her the weaker sex. It is true that up to about a century ago, women had a distinctly lower life expectancy than men, largely because of the deadly danger of childbearing. With the introduction of antiseptic techniques to the childbed, this danger largely passed and women now prove to have a life expectancy anywhere from three to seven years greater than men.

some 25,000,000,000,000 (that is, 25 trillion) erythrocytes, and the average woman would possess 17 trillion.

Once an erythrocyte has reached the stage where it has lost its nucleus, it can no longer grow and divide. It can only continue in its own person, for as long as it can last. This is not long—its life of bumping along blood vessels and, in particular, squeezing through the capillaries is a strenuous one. The average length of time an erythrocyte endures is 125 days. Signs of disintegrated erythrocytes that have reached the end of their useful and hard-working lives can be seen under the microscope as *hemoconia* (hee'moh-koh'nee-uh; "blood dust" G), or *blood dust* in direct English. This is filtered out in the spleen and absorbed there by large scavenging cells called *macrophages* (mak'roh-fay-jez; "large eaters" G).

On the average, then, we can count upon having 1/125 of our erythrocytes broken up each day, or 2,300,000 each second. Fortunately, the body is perfectly capable of making new erythrocytes at that rate continuously, throughout life, and at even faster rates, if warranted. One way of stimulating erythropoiesis is to have the blood continuously low in oxygen. This would be the case at high altitudes, where the air is thin. Under those circumstances, more erythrocytes are formed and, among people who live at high elevations, the erythrocyte count can be maintained at an 8,000,000 per cubic millimeter level.

In the larger blood vessels, the erythrocytes tend to stack themselves flat side to flat side. This is called "rouleaux formation" from the French, but we can picture it best, in less esoteric fashion, by thinking of a stack of poker chips. Blood flows through the larger vessels more easily when the erythrocytes neatly stack themselves in this fashion. However, rouleaux formation is not possible in the capillaries, which are in diameter barely, if at all, wider than the erythrocytes themselves. The erythrocytes must crawl along the capillaries singly and slowly, shouldering their way through the narrow openings rather like a man inching through a close-fitting tunnel on hands and knees. This is all for the best, since it gives them ample time to pick up oxygen, or to give it off.

A single erythrocyte contains about 270,000,000 hemoglobin molecules and each molecule possesses four heme groups. Each heme group is capable of attaching one oxygen molecule to itself. Therefore, an erythrocyte that enters the lung capillaries with no oxygen emerges at the other end with a load of something more than a billion oxygen molecules. The same volume of water by simple solution could not carry more than 1/70 the quantity. The existence of hemoglobin thus increases the efficiency of the bloodstream as an oxygen carrier some seventyfold. Instead of having a 4½-second reserve supply of oxygen in our bloodstream, we have a 5-minute supply. This is still not much, and a few minutes' asphyxiation is enough to throttle us, but at least it gives us a sufficient margin of safety to carry on life.

When oxygen diffuses across the triple barrier (the alveolar membrane, the wall of the capillary, and the membrane of the erythrocyte) and attaches itself to the hemoglobin molecule, a new compound—*oxyhemoglobin* —is formed. It is the oxyhemoglobin that is the bright red we actually think of as the color of blood. The hemoglobin itself, unoxygenated, is a bluish-purple color. As the blood passes through the systemic circulation and loses oxygen, it gradually darkens in color, until in the veins it is quite blue. You can, as a matter of fact, see this blue color in the veins on the back of your hand, on the inside of the wrist, and anywhere else that veins come near the surface of the body, if you are sufficiently fair-skinned. Sometimes the color appears greenish, because you see it through a layer of skin that may contain a small amount of a yellowish pigment. Nevertheless, few of us associate this blue or green color with blood, for the blood we actually see during bleeding is always bright red. Even if we cut a vein and allow the dark blood to gush outward, it would absorb oxygen as soon as it touched the air and it would turn bright red.

The bright red oxygenated blood is called *arterial blood,* because it is to be found in the aorta and in the other arteries of the systemic circulation. The dark unoxygenated blood is called *venous blood,* because it is to be found in the veins of the systemic circulation. This is not entirely a good set of names, for of course the situation is reversed in the pulmonary circulation. The pulmonary artery leads unoxygenated blood to the lungs and thus carries venous blood despite the fact that it is an artery. As for the pulmonary vein, it carries the freshest arterial blood in the body.

ANEMIA

Any deficiency in erythrocytes or hemoglobin (or both) is, of course, serious for the body economy. Such a condition is called *anemia* ("no blood" G, a slight exaggeration of the condition). In severe anemia the number of erythrocytes may sink to a third of the normal and the quantity of hemoglobin may sink to a tenth. In an anemic the efficiency of oxygen transport is reduced and the amount of energy available for use is correspondingly reduced. Consequently, one of the most prominent symptoms of anemia, aside from paleness, is the ease with which one tires.

The most direct cause of anemia is the loss of blood through a wound, either through accident or disease. Such blood flow is *hemorrhage* ("blood flow" G). In addition to the obvious external flows as a result of cuts, stabs, and scrapes, there is the possibility of *internal hemorrhage* as a result of physical injury or of a disease such as a bleeding ulcer. Nor need the hemorrhage be massive. A small but chronic hemorrhage, such as that involved in tuberculosis lesions of the lungs, can also induce a degree of anemia.

The danger of massive hemorrhages is twofold. First, a quantity of fluid is lost, and second, a proportional quantity of each of the chemical components of that fluid (of which the most important, next to water itself, is hemoglobin) is lost. The body has several devices for making up a reasonable loss of fluid. The arterioles contract, reducing the capacity of the circulatory system, so that the diminished quantity of fluid remaining in the body is nevertheless maintained at as nearly normal a blood pressure as possible. (Normal pressure is more important than normal volume.) The spleen also contracts, adding its reserve supply of blood to the general circulation. Fluid is withdrawn from the tissues outside the circulatory system and is added to the blood; the patient, of course, will also add fluid by drinking.

Where the loss is not too great, water is replaced quickly enough, but a longer time is required to replace some of the substances dissolved in the plasma, particularly the complex protein molecules. The erythrocytes are replaced most slowly, so that there is a definite period of *posthemorrhagic anemia*. This may last 6 to 8 weeks after the loss of a pint of blood, but it is not serious. The body has its margins of safety and a temporary mild anemia will not interfere with ordinary living. Thus, a person can give a pint of blood to the Red Cross without any subsequent discomfiture to speak of if he is in normal health to begin with.

Larger loss of blood is correspondingly more serious. If more than 40 per cent of the blood is lost, the body's machinery is incapable of bringing about reasonable recovery quickly enough for its needs. It is then advisable to transfer blood directly into the patient's bloodstream. The blood may come from the reserves of a blood bank or from a living donor, and the process is *transfusion* ("pour across" L).

Unfortunately, transfusion cannot involve any two persons taken at random. Human blood falls into four main types, which are categorized as O, A, B, and AB. The erythrocytes of a man with blood type A contain a substance we can call *A*, while those of a man with blood type B contain one we can call *B*. A man with blood type AB has erythrocytes containing both *A* and *B*, whereas one with blood type O has erythrocytes containing neither *A* nor *B*.

These groups are not of equal size. In the United States, out of every 18 people, 8 are of blood type O and 7 are of blood type A. Only 2 are of blood type B and only 1 of blood type AB.

It so happens that blood plasma may contain compounds capable of reacting with *A* or with *B*, causing erythrocytes containing the particular substance to clump together, or *agglutinate* ("glue together" L). We can call the substance causing *A* erythrocytes to agglutinate *anti-A*, and that causing *B* erythrocytes to agglutinate would be *anti-B*. A person of blood type A, with *A* in his erythrocytes invariably has anti-*B* in his plasma. (Naturally, he would not have anti-*A* for that would cause his own erythrocytes to agglutinate and kill him.)

In the same way, a person of blood type B would be expected to have anti-*A* in his plasma. A person of blood type AB, with both *A* and *B* in his erythrocytes would have to have neither anti-*A* nor anti-*B* in his plasma; a person of blood type O, with neither *A* nor *B* in his erythrocytes, would have both anti-*A* and anti-*B* in his plasma.

The situation will undoubtedly be made clearer by this brief table:

	Red cell	*Plasma*
Blood type O	—	Anti-*A*, anti-*B*
Blood type A	*A*	anti-*B*
Blood type B	*B*	anti-*A*
Blood type AB	*A, B*	—

Ideally, in making a transfusion, one should have both donor and recipient of the same blood group. Suppose that through accident or ignorance, though, the donor and the recipient are not of the same blood group. Suppose that blood is taken from a B-donor and given to an A-recipient. There are two possibilities of agglutination. First, the B donor has blood containing anti-*A* in the plasma. This anti-*A* agglutinates the erythrocytes of the A-recipient. This, however, is not usually very important. The quantity of anti-*A* in the plasma being transfused is not very much and what is present is quickly diluted by the larger quantity of blood in the recipient's own body. However, the second possibility is more serious. There is anti-*B* in the plasma of the A-recipient. The erythrocytes in the blood of the B-donor are themselves agglutinated by the large quantity of anti-*B* in the patient's blood supply. What the patient receives, then, is not real blood but a helping of clumped erythrocytes that promptly block his blood vessels, with frequently fatal results. The wrong transfusion is worse than none.

The important precaution is to prevent the donation of erythrocytes that will be agglutinated by the recipient's plasma. To illustrate, a patient with blood type A has anti-*B* in his plasma and therefore must not receive erythrocytes containing *B*. That eliminates donors with blood type B or AB. It does not eliminate a donor of either blood type A or blood type O. You can see for yourself, by the same line of reasoning, that a B-patient can receive blood from a donor of blood type B or blood type O, but not from one of blood type A or blood type AB. In fact, it is easy to prepare a table such as the following:

Patient	Possible donor
O	O
A	A, O
B	B, O
AB	AB, A, B, O

As you can see, an O-donor can be used for any patient; he is sometimes referred to as a "universal donor." (Actually, a particular O-donor may have enough anti-*A* or anti-*B* in his plasma to cause complications to A, B, or AB patients. It remains safest to use a donor of the patient's own blood group.)

Where large amounts of blood are lost, the fluid loss becomes more dangerous, from a short-term point of view, than the loss of erythrocytes. The fluid loss may be great enough so that despite all the body's compensatory mechanisms what is left is insufficient to maintain normal blood pressure, and this is more immediately perilous than hemorrhagic anemia.

In this case, transfusion must be completed without delay, and if blood is not available transfusion of plasma alone is far, far better than nothing. Blood pressure is maintained and the anemia can be dealt with in more leisurely fashion.

There are even advantages to making use of the plasma alone. Plasma keeps better than whole blood does. In fact, plasma can even be frozen and dried under vacuum to yield a powder that will keep indefinitely and that will be ready for use any time sterile distilled water is added. Secondly, with no red cells present, there is no need to worry about blood types and agglutination.

(There are, of course, blood types in existence other than A, B, O, and AB. In recent years the number of detectable types has increased greatly, but these other blood types are not usually of importance in transfusion.)

Even when there is no loss of blood itself, the body may fail to manufacture some essential component of the oxygen-carrying system. In that case, the body is subjected to the same weariness and disability (if not to the blood-pressure loss) that would result from hemorrhage. The most obvious anemia-producing failure of the body's chemical mechanisms is one where it does not produce hemoglobin, and here the weak point rests with the iron component. Except for the iron atoms, all portions of the homoglobin molecule can be manufactured from the ordinary and plentiful components of virtually all food. Nothing but serious and long continued undernourishment would interfere with the process of making hemoglobin protein; and this would interfere with protein manufacture generally and not with hemoglobin alone.

The iron atoms (four per molecule) make hemoglobin a special case. Food is not generally rich in available iron. Iron that forms part of an organic molecule, such as heme itself, is not easily absorbed by the body. For this reason meat and eggs, which are rich in iron, nevertheless offer only a marginal supply. Adult men need not be seriously concerned, since the body conserves its iron efficiently and loses virtually none of it, always barring hemorrhage. In the case of children, who grow and whose supply of hemoglobin must increase with the years, the available iron content of the food is a more crucial matter. The iron-enriched cereals constantly fed children these days are helpful in this respect.

Young adult women have a particular problem. Because of menstrual flow they are subjected to periodic losses of 25 milligrams of iron each month. This is not much in an absolute sense (about a thousandth of an ounce) but it must be made up, and young women, often intent on remaining slim at all costs, may not have sufficient iron in their limited diet. In any case, *iron-deficiency anemia* is more common among young women than among other classes of the population.

Fortunately, iron-deficiency anemia is easily treated by the addition of iron to the diet. Iron is most easily absorbed when it is in the form of inorganic salts and "iron pills" are frequently used during pregnancy. Although the menstrual losses come to an end then, the mother's body is depleted of its iron supply in order that the baby might be started with an oversupply. The baby's iron supply, after all, must be more than merely enough to meet its needs at the new-born moment. It must also be enough to allow for new blood manufacture during its first six months of life when milk, very poor in iron, is almost the only item on its diet.

Anemia may exist even when iron is plentiful provided the oxygen-carrying mechanism is disabled in some other fashion. Occasionally, for instance, there is a flaw in body-chemistry which brings about the manufacture of a hemoglobin very slightly different from ordinary hemoglobin. Such an *abnormal hemoglobin* is invariably less efficient as an oxygen-carrier.

The most common form of such an abnormal hemoglobin gives rise to *sickle-cell anemia*. This particular abnormal hemoglobin is less soluble than normal hemoglobin and actually precipitates out of solution within the erythrocyte when the oxygen content of blood is low, as in the veins. The erythrocyte, with the abnormal hemoglobin crystallizing within it, is distorted into bizarre shapes, sometimes into crescents resembling sickles (whence the name of the disease). The distorted erythrocytes are weakened and fragile, breaking up easily to produce an anemia. This is an inherited condition for which there is no treatment. It is found almost exclusively in natives of certain West African regions and in their descendants, including a number of American Negroes.

Striking more arbitrarily is a deficiency in the body's ability to make the material of the erythrocyte itself. This material is the *stroma* ("mattress" G; that is, something on which the cell contents can rest). The erythrocytes that are formed with defective stroma are abnormally weak and have an average life of 40 days rather than 125 days. The erythrocytes decrease steadily in number until a count of less than 2,500,000 per cubic millimeter may be reached. The individual cells are usually larger than normal but that does not compensate for the loss in numbers. This disease, *pernicious anemia* ("pernicious" meaning "deadly"), was so called because until the 1920's, no successful treatment was known and death was inevitable.

In the 1920's it was found that large quantities of liver in the diet would relieve the condition, and by the late 1940's small quantities of a compound called vitamin B_{12} or *cyanocobalamin* (sy'an-oh-koh-bawl'uh-min; thus named because its molecule contains a cyanide group, a cobalt atom and an amine group) were isolated from liver and found to be the curative principle. The vitamin can now be formed cheaply through controlled bacterial fermentation in the laboratory, and, at the cost of taking pills at periodic intervals, people with pernicious anemia can lead normal lives. The anemia is no longer pernicious.

Chemicals accidentally introduced into the body, or toxins from invading microorganisms, can in one way or another damage the oxygen-carrying capacity of the blood. The malarial parasite infests the erythrocyte and breaks it up. This is *hemolysis* (hee-mol'i-sis; "blood destruction" G), and it is during the period of hemolysis when the patient is attacked by the violent chills that accompany malaria. The venom of snakes and other creatures may also hemolyze the blood, or sometimes bring about erythrocyte agglutination. In either case, the results are often fatal.

The gas carbon monoxide represents a danger originating in the inanimate world. Like oxygen, carbon monoxide will combine with the hemoglobin in blood. Unlike oxygen, carbon monoxide is not freely given up again. It remains bound. Even when the air contains but small quantities of carbon monoxide, molecule after molecule of hemoglobin is tied up and rendered unavailable for oxygen transport.

Other gases can behave similarly, but carbon monoxide is most dangerous because it is most common. It can be formed in any poorly ventilated furnace, and it is also to be found in automobile exhaust and in one common form of cooking gas. All three sources of carbon monoxide are commonly implicated in death by asphyxiation, either as accident or suicide.

Too much of a good thing can also be deadly. The manufacture of erythrocytes is stimulated, as I said earlier, by a lower than normal content of oxygen in the blood. Ordinarily, as the number of erythrocytes in the blood is increased, the oxygen content also increases and the rapid formation of new erythrocytes levels off. It is possible, though, that the blood vessels feeding the bone marrow (where the erythrocytes are formed) are thickened, perhaps through atherosclerosis, so that the blood supply to those tissues is cut down. The marrow suffers from an oxygen shortage, which is the result of the narrowed blood vessels and not of any real lack of oxygen in the blood. The manufacture of erythrocytes is accelerated but does not correct the situation, so the manufacture continues without stint. The result is a dangerous crowding of the blood with erythrocytes, a condition called *polycythemia* (pol'ee-sy-thee'mee-uh; "many cells in the blood" G). The blood becomes thick and viscous at levels far beyond normal, circulation breaks down, and the results may be fatal.

LEUKOCYTES AND THROMBOCYTES

In addition to the erythrocytes, the blood contains normal cells, complete with nuclei. These are the *leukocytes* (lyoo'koh-sites; "white cells" G; since, unlike the erythrocytes, they lack pigment) and, as a matter of fact, they are often called simply *white cells*. They are also frequently termed *white blood corpuscles* by analogy with the red, though the leukocytes are not merely corpuscles but are true cells.

Most of the leukocytes, but not all, are manufactured in the bone marrow along with the erythrocytes. In the initial stages, the cells are first *myeloblasts* (my'uh-loh-blasts; "marrow buds" G) and then *myelocytes* ("marrow cells" G). They are formed in large quantities, but their

MYELOBLAST

EOSINOPHIL
MYELOCYTE

life is a hard and dangerous one and they do not generally live long. As a result, the number in the blood at any one time is merely 7000 per cubic millimeter, so that erythrocytes outnumber leukocytes some 650 to 1. Nevertheless, over the total blood supply, the number mounts up and the average man possesses some 75 billion leukocytes at any given moment.

The leukocytes exist in a number of varieties, differing among themselves in size and appearance. In general, they can be divided into two classes: those having a granular appearance (*granular leukocytes*) and those which do not (*agranular leukocytes*). The granular leukocytes usually have a nucleus of complicated shape, formed into two or more lobes. They are sometimes called *polymorphonuclear leukocytes* (pol'ee-mor'foh-nyoo'klee-ur; "nucleus of many forms" G) as a result. These ordinarily make up about two thirds of all the leukocytes in the bloodstream. They can in turn be divided into three types, depending on whether they stain with an acid dye such as eosin, a basic dye, or a neutral dye. The types are called, respectively, *eosinophils* (ee'oh-sin'oh-filz; "eosin-loving" G), *basophils* (bay'soh-filz; "base-loving" G), and *neutrophils* (nyoo'troh-filz; "neutral-loving" G). Of these the neutrophils are by far the most common.

As for the agranular leukocytes, which make up the remaining third of the white cells, they are characterized by large nuclei, simpler in shape than those of the granular leukocytes, without separate lobes and sometimes filling most of the cell. They are also divided into three types, which are, in order of decreasing size, *monocytes* (mon'oh-sites; "single cell" G, referring to the nucleus with its single lobe), *large lymphocytes,* and *small lymphocytes*. The small lymphocytes (I'll explain the name later) are not much larger than erythrocytes in size, and make up

BASOPHIL   NEUTROPHIL   MONOCYTE

about a quarter of all the leukocytes. Next to the neutrophils, the small lymphocytes are the most common of the leukocytes. The agranular leukocytes are not manufactured in the blood marrow—a subject I shall return to later.

It might be helpful to clarify matters by making a table:

	Number per cubic millimeter	
All leukocytes	7000	
Granular leukocytes	4625	
Neutrophils		4500
Eosinophils		100
Basophils		25
Agranular leukocytes	2375	
Small lymphocytes		1700
Monocytes		450
Large lymphocytes		225

The neutrophils are remarkable for their ability to progress through amoeboid motion. In fact they, and to a lesser extent other leukocytes, can pinch themselves enough to squeeze between the cells making up the thin walls of a capillary. In this way they can leave the circulatory system altogether and enter other tissues. This process is *diapedesis* (dy'uh-puh-dee'sis; "squeeze through" G).

This ability is extremely important, for the leukocytes are the shock troops of the body, being able to ingest and digest bacteria and other foreign particles. The invasion of bacteria at any point gives rise to a stimulus that brings about diapedesis nearby. Leukocytes carried to the spot by the bloodstream enter the tissue in quantity and devour the bacteria much as an amoeba might ingest some food particle. This process is *phagocytosis* (fag'oh-sy-to'sis; "eating of cells" G), and is one of the chief defenses of the body against infection.

(Another, more subtle, defense rests with certain proteins called *antibodies* in the blood plasma. These are manufactured under the stimulus of the presence of a foreign substance, let us say a bacterial toxin or the carbohydrate in the surface of a bacterial cell. The antibodies that are formed will combine specifically with the particular toxin or the particular bacterial cell surface that stimulated their formation. They will, in one way or another, put the toxin or bacterium out of action. In the course of a lifetime, an individual collects many types of antibodies, which by their mere existence and their continuous readiness to combine with invading substances make him resistant to numerous ills.)

The leukocytes do not find the bacteria harmless. The bacteria contain toxins that can in turn kill the leukocytes. Depending on the nature of the bacteria, a leukocyte may ingest as many as fifty or as few as two before being killed itself. At the site of infection, dead and disintegrating leukocytes collect as *pus* ("corruption" G). We are most commonly made aware of this when an infection of a hair follicle produces a boil. The blood collecting in that spot (bringing leukocytes to the scene of action) reddens and swells the area, while the fluid pressure makes it painful. The tissues between the focal point of infection and the skin are gradually broken down until finally only a thin membrane covers the conglomeration of pus and bacteria. The boil "comes to a head" and finally breaks, discharging the pus.

The number of leukocytes in the blood will rise or fall in response to certain abnormal conditions. A rise is called *leukocytosis* (lyoo'koh-sy-toh'sis; the suffix "osis" being used in medical terminology to signify a pathological increase of something) and a fall is *leukopenia* (lyoo'koh-pee'nee-uh; "poverty of white cells" G). The change is not necessarily uniform among the various types of cells and it is sometimes useful to make a differential count. To do this a smear of blood is stained and studied under the microscope. The changes in proportions of the various types can then be used as an aid in diagnosis. In acute infections, for instance, the neutrophil count goes up (*neutrophilia*).

A particularly serious type of increase in leukocyte number is the result of a cancer of the tissues producing them. Cancer is a disease characterized by unrestrained growth, and in this case the unrestrained growth consists of the manufacture of leukocytes without limit. The leukocyte count may go up to 250,000 per cubic millimeter, a rise of 35-fold, or even more. The white cells, by sheer quantity, invade and involve other organs, impairing their usefulness. The manufacture of erythrocytes is squeezed out by the overflow of the leukocyte-manufacturing mechanism, so an anemia is produced. This disease, invariably fatal, is *leukemia* (lyoo-kee'mee-uh; "white blood" G).

There is a third type of formed element in the blood, smaller and even less of a cell than is the erythrocyte. These are the *platelets,* so called from their platelike flatness. They are only half the diameter of the erythrocytes. Because they are involved in the phenomenon of clotting they are called *thrombocytes* (throm'boh-sites; "clotting cells" G). The thrombocytes are formed in bone marrow, as erythrocytes are. They begin as unusually large cells with a mass of multiple nuclei, called *megakaryocytes* (meg-uh-kar'ee-oh-sites; "giant nucleated cells" G). These are five times the diameter of an ordinary cell, twenty-five times the diameter of the final platelet. A week after formation, a megakaryocyte will develop graininess in its cytoplasm and will then fall apart into small pieces— the thrombocytes. The average lifetime of the thrombocytes once in the blood is only about 8 to 10 days. They are more numerous than leukocytes, but less numerous than erythrocytes; in numbers they come to 250,000 per cubic millimeter.

The thrombocytes come into play whenever, through some sort of wound, blood oozes out past the barrier of the skin. Contact with air breaks the thrombocyte, exposing its contents. These initiate a series of chemical changes which ends by converting *fibrinogen* (fy-brin'oh-jen; "fiber-producing" G), a soluble protein of the blood plasma, into the insoluble fibers of *fibrin* (fy'brin). The fibers of fibrin settle out of the blood as a fine network in which formed elements are trapped. The network plus its trapped cells make up a clot that seals off the wound and stops blood flow. With loss of blood prevented, the wound is repaired, and eventually the hardened clot, or scab, falls off.

Involved in the complicated process of clot formation are a number of *clotting factors,* all of which must respond properly before the final result is satisfactorily achieved. Sometimes one or another of the clotting factors may be missing and clotting fails to take place. This may be brought about deliberately. A sample of blood may be "defibrinated" by rapid stirring while it is being collected. The fibrin that is formed knots about the stirring rod and can be removed. If the formed elements are then centrifuged out, what is left is plasma minus the fibrinogen. This is *blood serum*. It does not offer the problem of clot formation and is so commonly used in laboratory manipulations that "serum" is almost a more familiar word than "plasma" in connection with blood.

Sometimes a chemical such as oxalate or citrate is added to blood as it is collected. This ties up the calcium ion in blood (one of the clotting factors) and clotting cannot take place. Sometimes a compound called *heparin* (hep'uh-rin; "liver" G, because it was first isolated from liver) is added to blood to prevent its clotting, for heparin ties up another of the clotting factors. This is handy during operations, when premature clotting is something a surgeon does not want to happen.

Unfortunately, it sometimes happens that a person is born with an inability to manufacture one of the clotting factors. When this happens, he is a "bleeder"; that is, once bleeding starts, it is extremely difficult to stop it. The condition is *hemophilia* (hee'moh-fil'ee-uh; "love of bleeding" G).

LYMPH

I have already pointed out that the circulatory system is not composed of truly closed tubes, since various leukocytes have no trouble escaping from the capillaries. It is not surprising that if large cells can do so the tiny molecules of water and of some of the substances it holds in solution can also do so. And, actually, under the pressure of the blood pumping through the arteries, fluid *is* squeezed out of the capillaries. This takes place at the arteriole end, where pressure in those tiny vessels is highest.

This exuded fluid, which bathes the cells of the body and acts as a kind of middleman between the blood and the cells, is called *interstitial fluid* (in'ter-stish'ul), because it is to be found in the interstices among the cells. In terms of sheer quantity there is far more interstitial fluid than there is blood plasma, eight liters of the former to only three of the latter in the average human being.

Interstitial fluid is not quite like plasma in composition, for not all the dissolved material in plasma manages to get through the capillary walls. About half the protein remains behind, so interstitial fluid is only 3 to 4 per cent protein, whereas plasma is some 7 per cent protein.

Naturally, the capillaries cannot continue to lose fluid indefinitely and there is a kind of circulation here, too. Some of the interstitial fluid re-enters the capillaries at the venule end, where the blood pressure is considerably less than at the arterial end. In addition, the tissue spaces contain thin-walled capillaries that come to blind ends, and through these capillaries some of the interstitial fluid drains off. Once in the tubes, the interstitial fluid is called *lymph* ("clear water" L); it certainly resembles clear water when compared with the viscous red blood. The vessels themselves are the *lymph capillaries,* or *lymphatics.*

The lymph capillaries join to form larger and still larger lymphatics and eventually join to form the right and left *lymphatic ducts.* These drain into the subclavian veins just behind the collarbone, and in this way lymph is restored to the blood vessels. The left lymphatic duct is the larger of the two, and the largest lymphatic in the body. It is usually called the *thoracic duct* because it passes up through the chest, or thorax, to its point of junction with the subclavian vein.

The lymph flows very slowly through the lymphatics. There is no pumping action and, as in the case of the veins, the chief motive force is the pressure of the muscles during the ordinary activity of the body. Again, as in the case of many of the veins, the lymphatics have one-way valves which see to it that the fluid flows only in the proper direction. Because of the slow flow through the lymphatics, very little of the interstitial fluid is restored to the circulation in that way as compared with the direct return through the venule ends of the capillaries. Nevertheless, the lymphatics act as a useful regulating factor, since flow increases and decreases with the fluid pressure within the tissues, and it acts to keep the pressure at a normal level. The effect of the regulation is most easily noted in its absence. When there is blockage of the lymphatics for any reason, fluid accumulates in tissues, bringing about *edema* (ee-dee′muh; "swelling" G), or *dropsy* ("water" G). The larvae of a tropical worm sometimes invade the body and succeed in blocking the lymphatic system, bringing about so exaggerated a swelling of the legs, for instance, as to give the disease the name of *elephantiasis* (el-e-fan-ty′i-sis).

More localized and temporary edematous conditions occur in the neighborhood of a mosquito bite or bee sting. Edema also accompanies certain allergic reactions, as in *hives*.

Scattered along the lymphatics at various places, particularly at the elbow, knee, armpit, and groin, are little beanlike masses into which several lymphatics enter and from which a larger one emerges. These were originally called *lymph glands* ("acorn" L) because there seemed a resemblance to an acorn. The word "gland" has a peculiar history. Because the lymph glands are small bits of tissue, other small bits of tissue came to be called glands even where there was no resemblance at all to an acorn. Then it was discovered that some of these glands secreted fluids of various sorts, some through a duct leading to the surface of the skin or into the intestines, and some directly into the bloodstream. This is why anatomists came to call any organ that formed a secretion a gland, even when it was quite large and unacornlike. The liver, for instance, a huge organ weighing three to four pounds, is called a gland because it secretes a fluid into the digestive canal.

From this new standpoint, the lymph glands—the original glands—were no longer glands, since they secreted no fluid. For this reason an alternate name, *lymph nodes* ("knot" L, because lymphatics seem to come together at those points to form a knotlike swelling), has gained popularity. It is at the lymph nodes that the agranular leukocytes are formed, and this is why two of the varieties are called *lymphocytes*. Although the lymph contains no erythrocytes or thrombocytes to speak of, it is rich in lymphocytes, and the nodes themselves are crammed with them.

The nodes thus form a second line of defense against infection behind the first line of the neutrophils that go swarming into the tissues immediately affected. Any bacteria or other foreign matter that has evaded or forced its way past the neutrophils and has entered the circulation will be filtered out at the nodes. There bacteria are killed and toxins neutralized (for another function of the nodes is the manufacture of the plasma proteins that form antibodies).

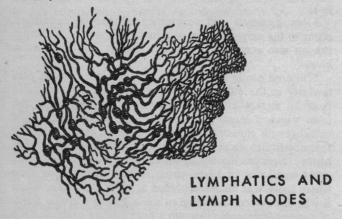

LYMPHATICS AND
LYMPH NODES

In the process, the nodes swell and may become painful; those being particularly affected which are nearer the original site of infection. The presence of "swollen glands" (as they are called by mother and doctor alike, in defiance of the newer terminology) at the angle of the jaw, in the armpit, or in the groin, is an indication of an infection of some sort.

The battle against infection is also carried on by larger bits of tissue similar in nature to the lymph nodes and therefore called *lymphoid tissue*. The spleen, which I mentioned earlier in the chapter, is the largest piece of lymphoid tissue in the body. It too is a filter, removing dying red cells and other debris. The macrophages that consume the debris are a form of monocyte. (This points up another function of the leukocytes, that of scavenging. The large leukocytes are useful since they can take bigger "bites.")

There are also scraps of lymphoid tissue in the throat and nose, standing guard, in a sense, at the points of greatest danger. These may be generally called *tonsils,* but the term is popularly restricted to the two rather large masses (an inch by half an inch) located where the pharynx meets the soft palate. There are also some 35 to 100 tiny scraps of lymphoid tissue at the rear of the tongue. These are the *lingual tonsils.* Where the pharynx meets the nasal passages there are a pair of *pharyngeal tonsils.* All these tonsils act as the lymph nodes do, in filtering out bacteria and combating them by lymphocytes. As in the case of the nodes, they can, when the struggle is hard, become inflamed, swollen, and painful (*tonsillitis*). In extreme cases, their protective function can be drowned out and they can, on the contrary, become a source of infection. It may then be advisable to remove them in a familiar operation referred to as a *tonsillectomy* (the suffix "ectomy," from a Greek word meaning "to cut out," is commonly used in medical terminology to signify the surgical removal of any part of the body. That tells you, for example, what an "appendectomy" is). The swelling of the pharyngeal tonsils may interfere with breathing if extreme enough, and they may be removed also. They are more familiarly known, by the way, as *adenoids* ("glandlike" G).

Certain primitive cells present in lymphoid tissue and in places such as lungs, liver, bone marrow, blood vessels, and connective tissue seem also to serve as scavengers, as do the macrophages in the spleen. They are sometimes lumped together as making up the *reticulo-endothelial system.* The "endothelium" refers to a layer of flat cells lining the interior of lymphatics and "reticulum" means a network. The reticulo-endothelial system is a network of cells that includes those lining the lymphatic vessels, in other words.

8

OUR INTESTINES

FOOD

Oxygen, taken by itself, is not a source of energy. To supply the energy required by the body, oxygen must be combined with carbon and hydrogen atoms making up the molecules to be found in food, forming carbon dioxide and water in the process. The ultimate source of food lies with the green plant. Green plants, at the expense of solar energy, combine carbon dioxide and water to form complex organic molecules formed largely of carbon and hydrogen atoms. These molecules fall mainly into three classes: carbohydrates, lipids (fats), and proteins. Any of these can be combined, through a complicated series of chemical reactions, with oxygen, and in the process the energy needed to support life is liberated.

Animals do not, as plants do, build up the complex molecules of carbohydrates, lipids, and proteins from the simple molecules of carbon dioxide and water and then live on them. Instead, they rifle the stores painfully built up by plants, or they eat animals that have already fed on plants.

The simplest method by which an animal can feed is exemplified in the protozoa. The amoeba, for instance, simply flows about an organism tinier than itself, or about a piece of organic matter, trapping it within itself in a water-filled *food vacuole* (vak'yoo-ole; "empty" L, because it looks like an empty hole in the cell substance, except for the particle of food which it contains). Into the vacuole are discharged specialized proteins called *enzymes* (en'zimez; "in yeast" G, because those found in yeast cells were among the first that were thoroughly studied). The enzymes act to accelerate the breakdown of complex molecules in the trapped food, forming simpler and smaller molecules which the cell can then incorporate into its own substance and build up into complex molecules somewhat different from those of the food and characteristic, instead, of the organism itself.*

This form of feeding requires that the article of food be smaller than the cell involved in the act; and as an organism grows larger, it would clearly be increasingly difficult to live on food particles that were always smaller than the size of its cells. It would be more efficient for one sizable organism to take as its prey another sizable organism, so that the predator could find its food supply in the form of a large and concentrated chunk. Of course, it could not very well engulf this large and concentrated chunk into any of its cells. Instead, it would have to break it down through enzyme action first and then absorb the products.

This would be all but impossible in the open ocean. As food was broken down, water currents would sweep the products away, were the process to take place in the open, and competing eaters would then take their share. The solution was to sequester a small portion of the ocean and in that portion to break down the food in peace and privacy.

The simplest organisms to accomplish this task were those ancestral to modern jellyfish. Essentially such organisms are a double layer of cells built in the shape of a hollow vase. The opening into the "vase" is the primitive *mouth* (AS). In the case of these animals the mouth is usually fringed by tentacles capable of stinging and stunning the prey, which could then be popped into the interior

---

* For some of the chemical details of the manner in which plants build up complex molecules and animals break them down, I refer you once again to my book *Life and Energy*. The same book considers the nature and workings of enzymes at some length.

of the "vase" or *gut* (AS). Into the gut would pour enzymes from the surrounding cells, and the complex organism would be broken down into simpler substances and as much as possible converted into soluble form. This process is *digestion* ("to dissolve" L). The dissolved materials produced by digestion are then absorbed into the various cells forming the lining of the gut, and those portions of the food incapable of being broken down ("indigestable residue") are cast out through the mouth opening. It is because of the development of the gut that the phylum to which the jellyfish belongs is called *Coelenterata* see-len'tuh-ray'tuh; "hollow gut" L). I mentioned this phylum in Chapter 1 as very probably ancestral to all other multicellular phyla.

Naturally, the larger the animal, the larger the gut, and the larger the individual piece of food that could be handled. Increased size invokes the square-cube law (mentioned in Chapter 5) and increased efficiency becomes also necessary. One obvious bit of room for improvement lies in the fact that the typical coelenterate has only one opening into its gut. Through that opening food must enter and indigestible residue depart. While one function operates, the other cannot.

The next step, then, carried through first by certain worms, was to add a second opening in the rear of the animal, a kind of back door. The original opening would remain for use as a food intake, the second would be reserved for residue ejection. The food would pass through the tube in one direction only and feeding could, in theory, be continuous.

All animals more complex than those worms (and this includes ourselves) retain the fundamental body plan of a tube with two openings running the length of the organism. The tube is called the *alimentary canal* ("food" L), the *digestive canal,* or simply the *food canal.* Substances within the alimentary canal are not truly within the body, therefore; merely inside a tube that is open to the outside world at both ends, like the hole in a doughnut. Actually this fact is not as obvious as it might be, since it's impractical to have the openings to the outer world really open; the canal would then be traversed by wind or current a bit too freely. Instead, both ends are usually pinched

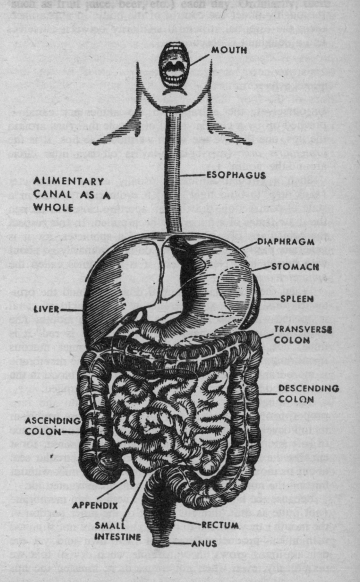

off, so that the volume between might be the more thoroughly under the control of the body. In appearance, then, the essential structural similarity between ourselves and a doughnut is obscured.

THE MOUTH

In ourselves, the entrance to the alimentary canal is pinched off by a circular band of muscle that runs around the lips, one that we use when we purse the lips. It is the *orbicularis oris* (awr-bi'kyoo-lay'ris oh'ris; "little circle around the mouth" L).

Such a circular muscle is usually called a *sphincter* (sfink'tur; "to bind tight" G); it ordinarily remains in a state of contraction, drawing an aperture closed as though the drawstrings of a purse had been pulled. In this respect the orbicularis oris is not much of a sphincter, for it is generally partly relaxed and we do not ordinarily go about with pursed lips. Nevertheless, it is sometimes called the *sphincter oris*.

When the lower jaw is pulled downward and the orbicularis oris relaxes, the entrance to the alimentary canal is opened wide and the portion we see is the mouth. The most obvious of its characteristics is that it is red. It is lined, not with skin but with a much thinner mucous membrane. Because it is thinner than skin, the membrane is more transparent and its color is that of the blood in the small blood vessels with which it is liberally supplied.

The mucous membrane folds outward into the face proper, forming a pair of *lips* (AS), a structure which, in its full development, exists in mammals only. The existence of soft and muscular lips in mammals makes sense, for it enables the mammalian infant to form a soft circular seal about its mother's nipple. It can then suck in milk without hurting the nipple and without sucking in unwanted air.

Because the lips are also covered with a thin membrane (not quite as thin, though, as that of the inner portion of the mouth) they are red in color. The lips are not supplied with mucus-producing glands of their own and yet the delicate lining grows uncomfortable when dry so that we periodically, even when not aware of it, moisten the lips with the tongue. In cold dry weather, when there is an increased reluctance to open the mouth, the membrane of the lips grows dry and may split, or "chap."

Although the mouth in the human being is of prime importance in speech (see Chapter 5) it serves, as it does in all other creatures beyond the very simplest, primarily as a food receptor. Where food is not already liquid or at least jelly-like in consistency, it must be made so, and for that purpose the double line of teeth rim the mouth. They are equipped for cutting, tearing, and grinding and their importance is plain enough to those who, through age or disease, have lost their teeth. Although modern dentistry has evolved excellent false teeth, none can quite serve as well as the genuine article.

If food particles are allowed to accumulate between the teeth, they will serve as breeding grounds for bacteria which will not only decay the teeth but will also inflame the fibrous tissues that cover the roots of the teeth. These tissues, covered with mucous membrane as is the rest of the mouth, are the *gums* ("jaw" AS). When inflamed they become uncomfortably tender and prone to bleeding, a condition called *gingivitis* (jin'ji-vy'tis; "gum inflammation" L). In extreme cases, pockets of decaying debris form between the tooth and the gum border, serving as a source of chronic infection, damaging the root and the neighboring jawbone and, eventually, leading to the loss of teeth. This condition is *pyorrhea* (py'uh-ree'uh; "flow of pus" G), and this has been blamed for most of the loss of teeth in people over thirty-five.

As food is chewed by the teeth, it is moved about by the nimble and muscular tongue, which sees to it that the food does not escape from between the teeth too soon. The lips and cheek stand guard along the outer rim of the teeth. The cheek muscles are especially useful here. The muscle is the *buccinator* (buk'si-nay'ter; "trumpeter" L, since stiff cheeks are essential to blowing a trumpet).

The coordinated movements of all these parts of the mouth are skillful enough to force chewing to completion without any portion of themselves being caught between the line of champing teeth. Once in a very rare while, coordination fails us, to our own great surprise, for the sharp pain of having bitten our own tongue while chewing is always accompanied by an almost involuntary feeling of incredulity.

Among animals the tongue has developed a variety of uses. It may be used to grasp food (as in the giraffe), to lap up liquids (as in cats), to sense the outer environment (as in snakes), to serve as a cooling agent (as in dogs), and even as an organ of attack (as in chameleons and toads) or entrapment (as in anteaters). In man, however, it has the unique function of making speech possible. The medieval punishment of cutting out the tongue did not rule out eating but did put an end to intelligible speech. The importance of the tongue in this respect is recognized in the fact that we will use the phrase "a foreign tongue" when we mean "a foreign language," to say nothing of the fact that "language" itself is derived from the French word for "tongue."

The tongue is covered with a series of small conical projections called *papillae*, which give the top of the tongue a rather velvety feel. (In the cat family they are large enough and hard enough to give the tongue something of the feel of a rasp, and anyone who has been licked by the family cat will know that on the whole it is not a comfortable sensation.) In among the papillae are small groups of cells which react to the chemical nature of the food that touches them and give rise to the sensation of taste. The cell groups are therefore the *taste buds*.

For some reason, not yet understood, deficiencies of various B vitamins produce as one of the prominent symptoms an inflammation of the tongue, or *glossitis* (glosy'tis; "tongue-inflammation" G). To give an example, *pellagra* (peh-lay'gruh; "dry skin" Italian) is a disease marked, as the name implies, by a roughened scaly skin. It is caused by a B-vitamin deficiency and is also characterized by a darkening and inflammation of the mucous membranes generally and of the tongue in particular. The same disease in dogs is called *blacktongue* because of that prominent symptom.

In chewing food, we do not merely break it up into smaller portions: we mix it with fluid and turn it into a soft, mushy mixture. The fluid used for the purpose is *saliva* (sa-ly'vuh) which is 97 to 99½ per cent water, but which also contains a mucopolysaccharide (see Chapter 3) called

mucin (myoo′sin), which even in small quantities suffices
to give saliva its stickiness and viscosity.

Saliva also contains an enzyme for which an old name is
ptyalin (ty′uh-lin; "saliva" G). This acts to bring about
the breakdown of starch to the simpler molecules of dextrins and sugars, so that digestion can be said to begin
in the mouth. In fact, if you begin with a mouthful of
potato, rich in starch and bland in taste, and chew it long
enough, you will begin to detect a certain faint sweetness
as sugar begins to accumulate. Since the Latin word for
starch is *amylum,* the enzyme is nowadays known as
salivary amylase (am′i-lays).

Saliva is secreted by three pairs of glands: the *sublingual
glands* (sub-ling′wool; "beneath the tongue" L), next the
submaxillary glands (sub-mak′si-leh-ree; "beneath the
jaw" L), and finally the *parotid glands* (puh-rot′id; "beside the ear" L). The names in each case are derived from
the position of the particular glands. The secretions of
these *salivary glands* (the general name for all three pairs)
are led into the mouth through narrow ducts.

We become most aware of the salivary glands when they
are infected by an all too common virus. The swelling that
results, of the parotid glands chiefly, is *mumps* (AS, possibly an old dialectical variation of what we would now call
"bumps" in reference to the swollen bumps just beside the
jaw). Most children get it at an early age. There are
advantages in this, for it is a mild disease and leads to
lifelong immunity; those unfortunate enough to escape
it as children may catch it as adults, at which time it may
be accompanied by unpleasant complications.

A human being will produce from half a liter to a liter
of saliva each day, and it is secreted even when we are
not engaged in chewing. It serves to keep the mouth
moist and clean and lubricates the tongue and cheeks so
that friction against the teeth in speaking or chewing does
not lead to irritation. The rate of salivary flow increases
quickly at the sight, smell, or even thought of food, particularly if one is hungry. The mouth will then "water."

THE STOMACH

Once the food is chewed and moistened and reduced to a semi-liquid state, the tongue rolls it into a ball (a mouthful of food is referred to as a *bolus,* in fact, which is just the Greek version of "ball") and pushes it backward into the pharynx. This is the act of *swallowing* (AS), or *deglutition* (dee-gloo-tish'un; "swallow down" L). At this stage the voluntary portion of the digestive process ceases. After that everything proceeds automatically.

The uvula moves up to close th nasal passages and the

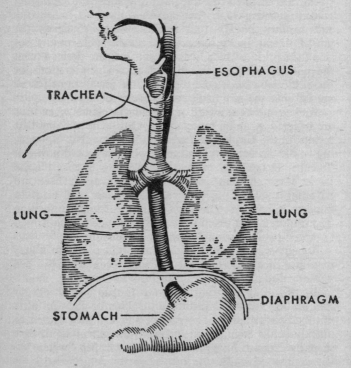

epiglottis covers the larynx. The food is debarred from returning by the tongue, and so it can move in only the one remaining direction, into a tube, leading downward, that lies just behind the trachea. This tube is the *gullet* ("throat" L), or *esophagus* (ee-sof'uh-gus; "to carry what is eaten" G). The esophagus is 9 or 10 inches long and is less than an inch wide. In its lower reaches, after traversing the thorax, it passes through the diaphragm (it and the aorta being the largest structures to do so). The esophagus has a muscular wall, with the muscle fibers arranged both longitudinally and in circles. The muscles in the upper few inches are striated, but they become smooth in the lower portion.

When food enters the esophagus it produces a dilation that stimulates a contraction of the circular muscles just above the area of dilation. This in turn causes a constriction of the ring of muscle just below, then just below that, and so progressively down the esophagus. Each muscle, after contracting, relaxes again and is then ready for another period of contraction when the next bolus comes by. The series of moving constrictions serves to push the food downward along the length of the esophagus. It might be supposed that gravity alone would accomplish this, but the bolus is still semisolid and viscous, and the muscular action, or *peristalsis* (pehr'i-stal'sis; "constriction around" G), serves to hasten matters. It is this peristaltic action that makes it possible to defy gravity and to swallow successfully even while standing on one's head; or, to bring in a matter now topical, under conditions of free fall. (Birds do not generally have this ability, and a bird drinking must throw its head up for each swallow in order to let gravity do its work.)

The esophagus, like the mouth and the alimentary canal generally, is lined with a mucous membrane. The slippery mucus protects the esophagus against the abrasive action of any still intact food particles in the bolus.

At the lower end of the esophagus, just below the diaphragm, the final circular muscle remains, usually, in a state of contraction, so that it serves as a sphincter. This is the *cardiac sphincter,* so named because it is located near the heart and *not* because it is part of the heart or has anything to do with the heart. At the approach of the bolus, the cardiac sphincter relaxes, the opening widens, and the food is shot into the stomach.

The stomach is the widest portion of the alimentary

canal and the most muscular. When empty, it is a rather
J-shaped tube, with its top pressed against the diaphragm.
The upper part of the stomach is usually puffed out a bit
with gas, even when the stomach is empty, and it therefore
bellies up above the cardiac sphincter. This upper portion
is the *fundus*. (Here is an example of a case where the
knowledge of Latin is a positive hindrance. The *fundus*
in Latin refers to the part of a container farthest from the
opening which, in the ordinary course of affairs, is the
bottom of the container. "Fundus" therefore has come to
mean "bottom." If the stomach were imagined as being
held by its entrance and its exit tubes, the fundus would
hang down at the bottom, too, but in the living man,
standing or sitting, the fundus of the stomach is at the top.)

The lower portion of the stomach is the *pylorus* (py-
law'rus; "gatekeeper" G, since it ends with a sphincter
that is virtually a gatekeeper for the rest of the canal).
The inner wall of the empty stomach exists in longitudinal
folds called *rugae* (roo'jee; "wrinkles" L), but as the
stomach fills with food, the rugae flatten out and disappear.
When full, the stomach becomes pear-shaped, being particularly distended in the fundus. Its capacity in an adult
is up to 1.5 liters. The capacity is much less in a youngster,
of course, and in a newborn baby it may only be 60 cubic
centimeters or so.

The stomach is the nearest thing in man to a storehouse
for food in the process of digestion. Not only is its capacity
larger than that of any other equivalent section of the
alimentary canal, but food usually remains in the stomach
three or four hours before passing on. This storage function
of the stomach is greatly exaggerated in grazing animals
that live on grass and other substances high in cellulose
content. Multicellular animals lack the enzymes to break
down cellulose into simpler substances that can be absorbed and used; cattle, for instance, are no exception.
The steer or cow depends upon the bacteria that infest its
alimentary canal to perform this task. The task takes time
and it is therefore necessary for the food to remain in the
stomach for an extended period, fermenting. The stomach
in cattle is for that reason extraordinarily large, holding
up to 300 liters, and is divided into four compartments.

In the two larger ones, the food is stored and there the bacteria do their work.

Initially cattle swallow grass hurriedly and with little chewing, but after this preliminary storage in the first two chambers of the stomach, they bring up the remains (now called "cud") and chew it thoroughly. When swallowed again, it passes into the final chambers of the stomach.

The lower reaches of our own alimentary canal are likewise filled with bacteria, which usually are not harmful and, at best, perform certain useful functions. Thus, they manufacture a number of vitamins in quantities greater than they themselves can use. We pick up the overflow as a kind of rent for their occupancy. Doubtless if we could store our food in our canal long enough, they would also break down cellulose for us, but we do not and, consequently, they do not. Cellulose passes through our own canal largely unchanged, so we cannot live on grass unless we run it through cattle first, then eat steak and drink milk.

While the food is stored in the stomach its walls undergo peristaltic contractions. With the sphincters at both ends of the stomach closed, the food must remain in the stomach, and the peristaltic action serves merely to mix it thoroughly with the digestive juices produced in that organ. Because of the gas which is usually trapped in the stomach, this churning of the food can produce gurgles sometimes heard as "stomach rumbling."

After the stomach has been empty for a period of time, the peristaltic contractions begin again, and the rumblings may then be louder. The gas, which now occupies the major portion of the volume within the stomach, is compressed in the process and its pressure against the stomach walls produces a feeling of pain, which we refer to as "hunger pangs."

The inner lining of the stomach is a mucous membrane into which are set numerous (up to 35,000,000) tiny glands that secrete a fluid called *gastric juice* ("stomach" G). This juice is unusual for a body fluid, in that it is strongly acid. It contains up to 0.5 per cent *hydrochloric acid*. This acid was discovered in the Middle Ages along with other strong mineral acids and was long considered a typical product of the inorganic world. Its existence among the delicate tissues of the body was first discovered in 1824 and it came as quite a shock to biologists.

Strong acid in itself accelerates the breakdown of proteins and carbohydrates into smaller substances; during early studies of the gastric juice it was felt that the hydrochloric acid was the digestive agent. Later a protein called *pepsin* ("digestion" G) was discovered and found to be far more effective than acid alone in breaking down proteins. It was one of the first enzymes discovered. A second enzyme, *rennin* ("to curdle" AS), acts upon milk in particular, curdling it as the protein content comes out of solution. The protein is "curds" and the clear liquid is "whey," and this (produced by adding a preparation of calf stomach to milk) is what Little Miss Muffet was eating, to the confusion of generations of youngsters who recite the verse without understanding it.

Thus it is chiefly protein that is broken down by the digestive processes of the stomach. To be sure, gastric juice does indeed contain an enzyme capable of breaking down fat, one called *lipase* (ly'pays; "fat" G). This enzyme, however, is a rather weak one, and even if it weren't, it would be incapable of acting under the acid conditions of the stomach. (Although the mechanism of the human body is marvelous beyond words and can never be sufficiently admired, it is not perfect. The existence of an enzyme under conditions that automatically make it useless is an example of this.)

The acid nature of the stomach contents has been made familiar to all of us by the frenzied activity of the advertising profession. The notion is often left with us that the stomach contents ought not to be acid at all, but of course the acidity of stomach juice is quite natural and beneficial. It is true that on occasion the acidity is higher than usual (*hyperacidity*) and that discomfort can then arise. Under such conditions, gas can accumulate and pressure on the walls of the stomach will increase. This gives rise to pain, as in the case of hunger pangs, but to pain that is more acute—enough, on occasion, to frighten a person into thinking something is wrong with his heart. (The stomach is located higher than most people realize. Ask someone to put his hand on his stomach and chances are he will put it on the navel or just above. The stomach is actually at the level of the lowermost ribs and its upper end is just below the heart.) Relief is usually brought about through the escape of some of the gas through the esophagus and mouth. At times the bubble of gas will carry some of the acid juices up into the esophagus in its struggle to escape. The stomach is insensitive to the acid, but the esophagus is not. The painful sensation in the chest that comes as a result is the familiar "heartburn."

A customary way of combating hyperacidity is to swallow a weak basic substance which will partly neutralize the acidity of the stomach. Sodium bicarbonate is the most familiar remedy. In neutralizing the acid it forms the gaseous carbon dioxide that swirls upward, collecting other droplets of gas with itself. When these escape relief usually follows.

It is not only gas that can escape the stomach. In extreme cases, the stomach can empty itself of its contents generally. This is the act of vomiting. In vomiting, the pyloric portion of the stomach contracts violently while the *pyloric sphincter* between itself and the lower regions of the alimentary canal remains firmly shut. The stomach contents can then only move upward through a suddenly relaxed cardiac sphincter. Vomiting can, of course, be useful, since it serves as a sort of natural stomach pump designed to empty the stomach of contents that may prove harmful if allowed to remain. There are drugs which will activate the vomiting reflex and will accelerate the stomach-emptying process. These are *emetics* ("vomit" G).

Quite extraneous sensations can produce the feeling of nausea that precedes vomiting. A steady rocking motion can do it, and to this some people are more susceptible than others. Those who have experienced "seasickness," the nausea resulting from the slow, relentless sway of a ship, well know the sensation of the utter worthlessness of life that can be produced.* The very word "nausea," by the way, comes from the Greek word for "ship."

Where vomiting is long continued, as in certain infections affecting the digestive system (intestinal flu, for

---

* There is the story about the seasick passenger who was assured jovially by the ship's steward that no one ever died of seasickness. "Please," muttered the passenger. "It's only the hope of dying that's keeping me alive."

instance, or similar diseases now dismissed by the physician with the awesome, all-inclusive term of "virus"), the condition can be debilitating and even dangerous. It is not merely the loss of food, but rather the loss of body fluid and the mineral ions it contains.

The resistance of the stomach wall itself to the strong acid its glands produce and to the meat-dissolving action of pepsin puzzled biologists for a long time. A piece of the stomach of another creature, used as food, is attacked and digested in the eater's stomach quickly enough. The answer to the paradox, apparently, is that in life the mucus secretions of the stomach (somewhat antacid in nature) coat and protect the wall.

This protection is not always perfect, particularly when the stomach is chronically hyperacidic, and even more so when the individual is particularly prone to tension and anxiety. Under those conditions, a portion of the stomach wall may be irritated and even eroded by the gastric juices to form an *ulcer* ("sore" L). Actually, the term "ulcer" can be applied to any break in the skin or mucous membrane which is accompanied by the destruction of tissue and the discharge of fluid, but in common speech the term is confined almost entirely to sores of the lining of the alimentary canal. One that is in the stomach is sometimes specified as a *gastric ulcer*.

There are cases, too, in which the stomach secretions contain virtually no hydrochloric acid. This condition is *achlorhydria* (ay-klawr-hy'dree-uh; "no hydrochloric acid" G). This is not necessarily serious, for though it decreases the efficiency of digestion in the stomach, the body can make out with the remainder of the alimentary canal. People with pernicious anemia, however, are invariably achlorhydric; before the lack of acid is dismissed as unimportant, the possibility of pernicious anemia must be considered.

THE PANCREAS AND LIVER

While food is in the stomach, the pyloric sphincter remains closed as long as any significant portion of the food remains not quite liquid. Gradually, however, the effect of gastric digestion and the addition of the gastric juices reduces the food to an entirely fluid condition. It is then *chyme* (kime; "juice" G). Only then does the pyloric sphincter relax so that the chyme, driven by peristalsis, passes in spurts into the next section of the canal. The chyme is virtually sterile, because stomach acidity has killed any bacteria originally present in the food (but bacteria will reappear and wax numerous in the lower reaches of the canal).

The chyme, on leaving the stomach, enters the *intestines* ("internal" L). These are also sometimes referred to as *bowels* ("sausage" L), because the long, flexible tube, with the periodic contractions produced in it by peristaltic action, bears a resemblance to a string of sausages. The intestines are divided into two portions, a relatively long section called the *small intestine,* which comes first, followed by a relatively short section called the *large intestine.* The "small" and "large" refers to the width, rather than the length. The small intestine is only 1½ to 2 inches in diameter at the point where it leaves the stomach and it narrows somewhat thereafter. The large intestine is up to 2½ inches wide.

The small intestine is 20 feet long or more (in a dead man, anyway; it may be contracted and rather shorter during life). In order to fit into the abdominal cavity, it coils intricately upon itself and, even so, succeeds in filling the major portion of the cavity. This length of intestine is necessary, for it is here that the main task of digestion is accomplished. The small amount of carbohydrate digestion in the mouth and the rather larger amount of protein digestion in the stomach can, in a pinch, be dispensed with. In fact, when for any reason it is necessary to remove part or even all the stomach (*gastrectomy*), it remains possible for the patient to lead a reasonably normal life. Without the storage capacity of the stomach, he must eat smaller and more frequent meals, but this might be a good idea, anyway.

Digestion in the mouth and stomach does not in any case proceed to completion. That is, although some of the foodstuffs are broken down to simpler compounds, the products are still not simple enough to be absorbed. (Small inorganic ions may be absorbed however. Cyanide ion can even be absorbed in the mouth—with drastic results, since cyanide ion is a deadly poison.) In the small intestine digestion does proceed to completion and it is there, particularly in the lower half, that absorption takes place. This is another reason for the length of the small intestine, since the entire length is needed to make sure that absorption is reasonably complete.

(Herbivorous animals, with a difficult-to-digest food supply, have proportionately longer intestines. The small intestine of a cow may be up to a hundred feet in length. Carnivorous animals have proportionately shorter ones. Man, who will eat anything and everything, is intermediate in this respect.)

The first ten or eleven inches of the small intestine is the *duodenum* (doo′oh-dee′num; "twelve" L). How the name came to be applied to a length of intestine shorter than twelve inches becomes clearer when you understand that the length was originally measured not in inches but in finger widths. In German, in fact, the section is known as *Zwölffingerdarm,* meaning "twelve-finger bowel." The duodenum is the section of the small intestine that receives the initial shock of the highly acid chyme as it pours through the pyloric sphincter, and its role is to neutralize that acid. For this reason, antacid secretions pour into the duodenum from two large glands. Despite this, the lining of the duodenum, like that of the stomach, is in constant danger, and *duodenal ulcers* may result. Since both gastric ulcers and duodenal ulcers are caused by the acid, pepsin-containing stomach secretions, they are sometimes lumped together as *peptic ulcers.*

Of the two glands whose secretions neutralize the acid of the chyme, the smaller is the *pancreas* (pan′kree-is; "all meat" G, because it lacks bones and fat and, as an animal organ, is completely edible without trimming). It is the second largest gland in the body, reddish in color, 5 or 6 inches long and rather carrot-shaped. It weighs about 3 ounces, and lies along the back wall of the abdomen behind the lower portion of the stomach. The wide end of its carrot shape snuggles up against the curve of the duodenum.

The *pancreatic juice* passes through a duct that opens into the duodenum about an inch and a half below the

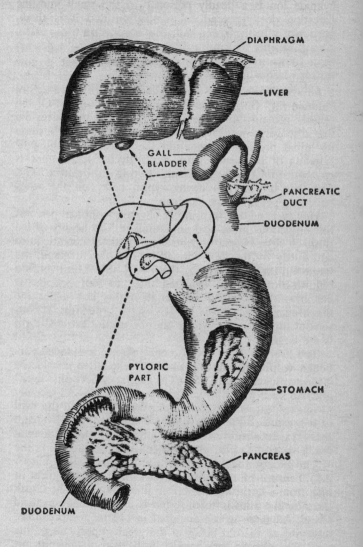

DIAPHRAGM

LIVER

GALL BLADDER

PANCREATIC DUCT

DUODENUM

PYLORIC PART

STOMACH

PANCREAS

DUODENUM

pyloric sphincter and through it about 0.7 liters of fluid are delivered each day. The pancreatic juice contains a large variety of enzymes, with one or more suitable for attacking each different type of foodstuff. There is a starch-splitting enzyme something like that in saliva. It used to be called *amylopsin* (am'i-lop'sin) but is now generally spoken of as *pancreatic amylase.* There is also a fat-splitting enzyme once called *steapsin* (stee-ap'sin; "animal fat" G) but now called *pancreatic lipase.* Of the several protein-splitting enzymes in pancreatic juice, the first discovered was *trypsin* (trip'sin; "to rub" G, because its initial preparation required the thorough rubbing or grinding of pancreatic tissue with glycerol). Another is *chymotrypsin* (ky'moh-trip'sin, "trypsin in the juice" G). Together trypsin and chymotrypsin carry on the work begun by pepsin in the stomach.

However, neither trypsin nor chymotrypsin can perform their functions in an acid environment. The acidity of the stomach chyme is partly neutralized by the pancreatic juice and partly by the fluid formed by the second gland, the largest in the body. This is the *liver* (AS). The liver is a reddish-brown organ weighing three to four pounds. It lies over the stomach and to its right, just under the diaphragm, and is partly hidden by the lower ribs. It consists of four lobes, of which the one on the extreme right is the largest.

The liver, perhaps because of its size, commanded particular respect among the ancients, who often considered it the particular seat of life. (The similarity between "liver" and "to live" may not be entirely accidental.) As recently as Shakespeare's time, common expressions used the liver as a symbol of the state of the emotions. "Lily-livered," that is, a liver poorly supplied with blood, was a synonym for "coward," to give the most common example.

The secretion formed by the liver is called *gall* ("yellow" AS) because of the yellowish cast of the fresh juice, or *bile,* from a Latin word of uncertain derivation. The Greek term for the juice is "chole" and is used in many medical terms. All three words are used in common expressions, testifying to the old belief in the great importance of the fluid. The Greeks considered it to be two of the four important fluids of the body, for they felt it consisted of two varieties, one colored black and one yellow. (This is not so; there is only one bile, though it may be differently colored, depending on the state of its freshness.) The Greeks felt a person who suffered from an overproduction of black bile was "melancholic," and one suffering an overproduction of yellow bile was "choleric." The supposed connection of the liver and emotion is clear, for we still use those words to indicate dispositions that are given to sadness or anger, respectively. We also speak of "gall" when we mean "impudence" and sometimes use "bilious" as a synonym for "choleric."

The bile is conducted toward the duodenum by means of the *hepatic duct* (hee-pat'ik; "liver" G) and the average output is about 0.5 liters a day. Bile is secreted continuously, but between meals a quantity of it is stored in a special sac called the *gall bladder* (AS; the term "bladder" can be used for any distensible sac). The gall bladder lies just under the right lobe of the liver and is a pear-shaped organ about two or three inches long. It has a capacity of only about 50 milliliters, but water is reabsorbed by the liver from the gall bladder so that the bile stored there is concentrated ten- or twelve-fold. The gall bladder therefore holds the dissolved substances of some 600 milliliters of ordinary bile or, roughly, that of a day's supply. When food arrives in the duodenum, the muscular wall of the gall bladder contracts and its contents are forced through its *cystic duct* (sis'tik; "bladder" G), which joins the hepatic duct to form the *common bile duct*. The common bile duct joins the pancreatic duct just before entrance to the duodenum, which thus receives a mixture of ordinary bile, concentrated bile, and pancreatic juice.

The concentration of the bile in the gall bladder is not without its dangers. The bile is rich in cholesterol (a fatty substance with an ill fame nowadays as the substance characteristically deposited on the inner walls of arteries in atherosclerosis). Cholesterol is rather insoluble, and since the bile is concentrated it may happen that small crystals of cholesterol will precipitate. In some people more than in others (for a reason not as yet known) there is a tendency for these crystals to aggregate into a sizable *gallstone*. If these grow large enough to block the cystic duct and interfere with the flow of bile, they can give rise to considerable pain. Sometimes the solution to the problem is the removal of the gall bladder altogether. This does not appear to interfere seriously with the patient, and it does free him of pain.

The bile does not contain any enzymes, but it is important to digestion anyway. It contains compounds (*bile salts*) possessing a detergent action. This encourages fat to break up into small globules that mix more or less permanently with water; that is, an *emulsion* ("to milk out" L) after the fashion of homogenized milk. This is important, because enzymes are as a class water-soluble and can only act upon substances that are dissolved in water or are, at the very least, well mixed with it. Pancreatic lipase, in its attack upon fat, could at best only perform its work at the edges of fat droplets where in contact with water. If these droplets are large, the fat in the interior would remain unaffected; only with very small, bile-induced droplets does fat digestion become efficient. Without the flow of bile, then, most of the fat in the diet would remain undigested.

Bile contains a number of waste products the body gets rid of via the liver and the alimentary canal. It is in the liver, for instance, that hemoglobin molecules are broken down after the natural disruption of aged erythrocytes. The heme (the portion of the molecule containing the iron and consisting of four circles of atoms set in a larger circle) is broken away from the protein portion of the molecule. The large circle of atoms is then broken and the iron atom is removed. What remains is *bile pigment,* so called because these waste remnants of heme are variously colored red, orange, and green and lend the overall yellow-green color to bile. Through its further course in the intestines the food retains the color of bile pigments, and when finally eliminated from the body it is still reddish brown in color.

A small quantity of bile pigment is absorbed into the blood and eliminated in the urine. This lends both blood plasma and urine its light straw or amber color. Under certain conditions an abnormally high quantity of bile pigment gets into the blood. Sometimes it is because erythrocytes are being broken down at an unusually high rate, and the formation of bile pigment rises. Or else the bile duct is obstructed so that bile pigment cannot be eliminated in the usual manner and must find its way into the blood instead. Whatever the reason, the yellow-green of the bile pigment then shows up in the skin and mucous membranes, in the whites of the eyes, too, and gives a person a sickly and unpleasant yellowish cast. The condition is known as *jaundice,* from the French word for "yellow."

The liver is the chief chemical factory of the body with a capacity for conducting an amazing array of chemical reactions. Cholesterol and bile pigments are not its only excretory products. Any chemical that gets into the body and cannot be broken down for energy or incorporated into the body structure is likely to end up in the liver for *detoxication* ("unpoisoning" G). The liver usually does this by adding some substance to the chemical which increases its solubility and hastens its elimination by way of the urine. This, of course, renders the liver liable to damage if the supply of foreign chemical is too great to be easily handled. The vapors of carbon tetrachloride, for one (which is used as a nonflammable drycleaner), or of chloroform (sometimes used as an anesthetic) may seriously damage the liver.

When liver tissue is damaged or destroyed, the active liver cells are replaced by fat and connective tissue. This gives the liver a yellowish appearance in place of the original reddish brown. This condition is *cirrhosis* (si-roh'sis; "tawny" G). Such destruction can have a number of reasons but is often associated with overindulgence in alcohol. Presumably, the liver in its continued attempt to deal with the alcohol is slowly and irrevocably ruined.

ABSORPTION

The chyme, with its admixture of pancreatic juice and bile, travels on out of the duodenum and into the small intestine proper, urged on by peristaltic action. The main body of the small intestine is divided, rather arbitrarily, into two sections. The first, making up about two fifths of the length, is the *jejunum* (jee-joo'num; "empty" L, so called because it is usually found empty in cadavers). The final three fifths is the *ileum* (il'ee-um), possibly called this from a Greek word meaning "twisted" because of the coils into which it falls.

Set all along the inner lining of the small intestine are numerous tiny projections something like those of a very fine Turkish towel. These projections are called *villi* (vil'eye; "a tuft of hair" L) and give the inner lining a velvety appearance. Their existence vastly extends the surface area of the intestine and facilitates absorption. In addition, by moving about ceaselessly, so that the liquid in the immediate neighborhood is kept constantly churned up, they further speed absorption. At the base of each villus is a group of cells that secrete still another fluid into the alimentary canal. These are the *intestinal glands,* and the fluid they secrete is *intestinal juice.* The fluid is also called *succus entericus* (suk'us en-ter'i-kus), which is simply "intestinal juice" in Latin.

The intestinal juice possesses a number of enzymes designed to break down the products produced by the digestive processes that have already taken place. Even the enzymes of the pancreatic juice do not complete the task of digestion to the point where the products are simple enough for absorption. But now the job is brought to an end. Thus, intestinal juice contains a number of *peptidases* (pep'tih-day'siz), enzymes designed to break down protein fragments (*peptides*) left behind by pepsin and trypsin. The fragments are broken down to the ultimate protein building blocks, the amino acids.

The carbohydrates are broken down in the mouth and duodenum (by salivary amylase and pancreatic amylase) to a simple compound called *maltose.* In the intestinal juice is an enzyme, *maltase,* which breaks each molecule of maltose into halves, the still simpler product being *glucose.* Another enzyme, *sucrase,* breaks molecules of ordinary table sugar (*sucrose*) into halves, forming glucose and *fructose,* and a third enzyme, *lactase,* breaks the molecules of the sugar in milk (*lactose*) into glucose and *galactose.* There is even a weak lipase present in case anything has been left undone by pancreatic lipase, and in this way fat molecules are broken down to *glycerol* and *fatty acids.*

The overall action of digestion can be summarized in the following table:

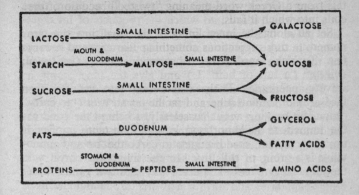

The substances shown on the far right are capable of being absorbed across the surface of the small intestine, and it is when this happens that food can finally be said to enter the body.

Within the body these simple molecules serve as building blocks for larger and far more complicated ones. They can be built up into carbohydrates, fat, and protein of the human type (*assimilation*), which means that they are put together into different combinations and arrangements than had existed in the original organism or organisms from which the food had been derived. (All living creatures on earth, however diverse in appearance, from the minutest virus to the largest whale, are made up of compounds that contain the same building blocks. This means that any one type of organism can serve as food for any other type, directly or indirectly. In the gastronomic sense, all life is definitely and incontrovertibly one.)

The same simple molecules that can be assimilated into tissue structure can also be combined with oxygen to form, eventually, water and carbon dioxide, plus nitrogen-containing wastes. When this happens, energy is released which can be used to power muscular activity and a variety of other energy-consuming tasks of the body (including assimilation, which requires an input of chemical energy).

Absorption through the intestines is similar in design to absorption through the lungs. Within each villus is a network of capillaries into which the end products of protein and carbohydrate digestion are passed. In addition, there is a lymph capillary into which the products of fat digestion move. The droplets of fatty materials turn the clear lymph in those capillaries milky in appearance and because of this these capillaries are termed *lacteals* (lak'tee-ul; "milk" L).

The capillaries of the villi collect into venules, then veins, and finally discharge into the *portal vein* ("to carry" L), a wide, short vein that starts just behind the pancreas and moves to the liver. Its function, as the name implies, is to carry the collected products of carbohydrate and amino-acid digestion to the liver. In the liver, the portal vein breaks up into vessels (*sinusoids*) that are rather wider than capillaries and form a network throughout the liver. Along the walls of the sinusoids are *Kupffer's cells* (named for the German anatomist who discovered them in the mid-19th century). These are members of the reticulo-endothelial system and act as scavengers, filtering out of the blood any debris, particularly any bacteria, that may have made their way across the intestinal wall.

In addition, the cells bordering the sinusoids strip the blood of its excessive supply of glucose and amino acids. The glucose is put together to form large molecules of a kind of starch called *glycogen* (gly'koh-jen; "to give rise to sugar" G, since it can be broken down to sugar again at need). Any fructose or galactose present is converted first to glucose and then to glycogen. The glycogen remains behind in the liver, and the blood emerging from it contains only a small and fixed amount of glucose. The amino acids are taken up by the liver cells and combined into protein molecules and the blood emerging from the liver has the proper content of those, too.

Once past the liver, the blood in the sinusoids is collected again into the *hepatic vein,* which empties into the inferior vena cava and thereby enters the general circulation. As the blood passes through the capillaries of the tissues and as plasma filters across the capillary walls and becomes interstitial fluid, the various cells of the body absorb the glucose fed into the blood by the liver and break it down for energy. They also absorb the proteins, pull them apart, and build up their own special varieties.

Between meals, when the requirements of the cells place

a drain upon the glucose supply that is not made up for immediately by new glucose from the intestines, the liver draws upon its reserve supply of glycogen. The glycogen, stored in the lush times just after a meal, is now broken down to glucose and fed, little by little, into the bloodstream.*

The body, within certain limits, can convert one form of foodstuff into another. To give an example: the liver can only store enough glycogen to see an individual through about 18 sedentary hours. If glucose continues to reach the liver, after the liver is completely gorged with glycogen (as for well-nourished individuals it too often is) the glucose must be converted to fat and stored in that form. Fat is a more concentrated form of chemical energy than glycogen is, and it can be stored in indefinite amounts. Contrariwise, when as a result of a day or more of fasting, the liver's glycogen supply is gone, the body can call upon its fat stores as a source of blood glucose.

Fat is stored in the cells of a type of connective tissue. The fat collects as droplets within the cells until these become little more than tiny goblets of fat surrounded by a thin rim of protoplasm. Collections of such cells form *adipose tissue* ("fat" L).

Adipose tissue is a normal component of the body, making up some 15 per cent or even more of the weight of an average, not particularly fat, person. (To be sure, in fat people, the percentage of adipose tissue can easily rise to where it makes up more than half the total weight of the body.) It is only as a result of lengthy starvation that a body is depleted of its fat stores and the victim then takes on an emaciated appearance that departs as far from our normal standards of "good looks" as does the appearance of an obese man.

The normal quantity of adipose tissue in an individual is enough to keep him going for a month without eating, provided he is plentifully supplied with water—and with vitamin and mineral pills. Adipose tissue supplies only calories, and it is quite possible that the body's reserve store of trace components, such as the vitamins and min-

* All these changes which I describe so quickly and, indeed, negligently, are actually very complicated. My book *Life and Energy* will give you some idea of these complications.

erals, is dangerously low regardless of how well supplied it may be with calories alone.

Fat men can have as much as a year's supply of calories tucked here and here about the body, but this does not mean they can fast for a year, even with water and trace components (including minimal protein) supplied. For many reasons, some physical and some psychological, men grow hungry and crave food even when their fat supplies are more than ample. It is for this reason—at least in part —that loss of weight through dieting is so long and torturing a procedure, and so often a failure.

Adipose tissue serves purposes other than that of a food store. For instance, it is a poor conductor of heat; when stored in the *subcutaneous layer* (sub-kyoo-tay'nee-us; "beneath the skin" L, which exactly describes its location), it acts as an insulating blanket against the cold. Thanks to our lack of any real coating of hair, this is insufficient protection outside the tropic zone and clothing is required in addition. However, a whale, surrounded always by water at near-freezing temperature, and without a hair coating, makes out with a layer of fat ("blubber") up to six inches thick. Other warm-blooded creatures that spend much of their time in the sea are likewise well supplied with subcutaneous fat.

As it happens, there is more subcutaneous fat in the female than in the male and it is more evenly distributed. Women may perhaps feel a trifle annoyed at being told they are fattier (not fatter, but fattier, if you see the difference) than men, but it is this even subcutaneous layer of fat that softens and curves their outline—a consequence which, I have every reason to believe, is quite satisfactory to one and all. What may be the most irritating aspect of this phenomenon, as far as women are concerned, is that additional subcutaneous fat is stored, to a large extent, in the buttocks. This is carried to an amazing extreme among certain groups like the Hottentots of southern Africa. Hottentot women may develop rumps resembling the hump of a camel, with a similar structure and purpose. Such a development, called *steatopygia* (stee'a-toh-py'jee-uh; "fat rump" G), is useful for seeing a woman through a hard winter and is undoubtedly attractive to Hottentot men.

Beauty, as we too infrequently remember, is in the eye of the beholder.

Adipose tissue also has a protective function. It serves as a pad that cushions blows, and it is particularly useful as a cushioning support for organs such as the kidneys.

Fat is also stored to an unusual extent in the *omentum* (oh-men'tum, a word of uncertain derivation). This is a membranous sac enclosing the stomach. The *lesser omentum* lines the stomach on the liver side; the *greater omentum* lines the other side of the stomach and drapes down the abdominal wall and over the intestines like an apron. It is in this greater omentum that fat is stored and the excess fat here contributes to the "potbelly" affecting many middle-aged individuals.

The intestines themselves, by the way, are enclosed by a double membranous sac, the *peritoneum* (per'i-toh-nee'um; "stretched around" G). This serves to wall off the viscera from infection, so that the intestines resemble, in a way, the supermarket items that are "wrapped in plastic for your better protection." The peritoneum itself can, of course, be infected (*peritonitis*), with serious results. In the days before modern surgery, any wound or incision in the abdominal regions was almost bound to lead to peritonitis. Modern aseptic procedures have reduced the danger, and the use of modern antibiotics to battle the inflammation, should it arise, has helped further.

THE COLON

It takes food about three hours to pass through the yards of small intestine, and when it does so it finds itself at the entranceway to the last major portion of the alimentary canal, the large intestine. The large intestine is about five feet long and is often called the *colon* (koh'lon), which is the Greek name for it (and which is why we speak of inflammation of the large intestine as *colitis,* and of the pain caused by gas-distention of the intestine as *colic*). The main portion of the large intestine is divided into three sizable regions, depending upon the direction of the flow of its contents. The small intestine enters the large at the lower right-hand side of the body, near the groin. From that point, the large intestine leads upward to the bottom of the rib cage along the right side, and that portion is called the *ascending colon*. The large intestine then makes a right-angle turn to the left, leading under the liver, stomach, and pancreas, this section being the *transverse colon*. It then leads downward again along the left side to the hipbone, and that part is the *descending colon*.

The point of junction between the small intestine and the large is the *ileocolic sphincter* (il'ee-oh-kol'ik; that is, the sphincter between the ileum and the colon). This is not set at the very bottom of the ascending colon. Rather, it is

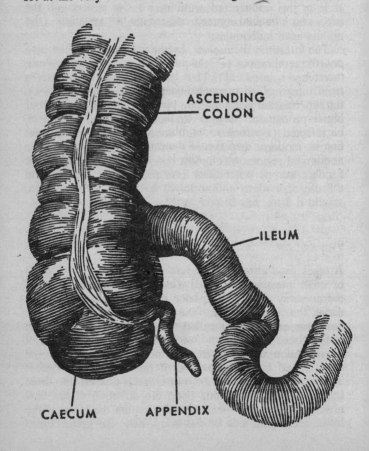

ASCENDING COLON

ILEUM

CAECUM APPENDIX

set about 2½ inches above the bottom, so that the lower-most bit of the ascending colon forms a dead end, or blind alley, and is called the *caecum* (see'kum; "blind" L). In the caecum, through the action of gravity, material can collect. In many herbivorous animals, such as the rabbit, this is exactly what is desired and the caecum is enlarged to the point where it makes up half the total length of the large intestine. It serves as a storehouse in which fermenta-tion can continue to take place.

In man, what is left of the caecum (a vestige perhaps of herbivorous ancestors) is of no particular use and can actually be a source of trouble. To the bottom of the caecum is attached a small appendage, or appendix (the two are essentially the same word), which is a further remnant of a once sizable and usable caecum. This ap-pendage is about 2 to 4 inches long and is shaped very much like a worm. In fact it is called the *vermiform ap-pendix* (vur'mi-form; "worm-shaped" L). An insignificant foreign body, an orange-pit, perhaps, that has survived digestion, can find its way by ill chance into this narrow blind alley set into a wider blind alley. This may set up first an irritation and then a dangerous inflammation (*ap-pendicitis*). It is only in the last century that removing the appendix in such cases (*appendectomy*) has become a simple operation without much danger of peritonitis.

There is no digestion to speak of in the large intestine. Digestion is finished. Absorption goes on, however, par-ticularly the absorption of water. The body has been lavish in its use of water in the various digestive secretions and to lose that water altogether would be very undesirable. As water is subtracted, the contents of the large intestine become increasingly solid. By the time the lower section of the descending colon has been reached, the contents are distinctly solid, although of course still soft.

At the bottom of the descending colon, the large in-testine makes an S-shaped curve to get to the center of the hip region and this short portion is the *sigmoid colon* ("S-shaped" G). The final four to five inches of the colon is vertical and this is the *rectum* ("upright" L). The open-ing of the rectum into the outer world is the *anus* (ay'nus), a word of doubtful origin, although some think it comes from the Latin word for "ring." Under ordinary circumstances, the anus is pinched off by a sphincter, the *anal sphincter,* or *sphincter ani.* (Actually, there are two of these, the inner one being some way up the rectum.)

The solid contents of the final portion of the alimentary canal make up the *feces* (fee'seez; "dregs" L), which consist in part of the indigestible residue of food, of fragments of cellulose and similar substances, of collagen and other constituents of connective tissue, all this being lumped under the name of "roughage" for obvious reasons. The feces also contain many bacteria, which proliferate enormously during the passage through the large intestines. Most of these bacteria are harmless, but some can be definitely dangerous. Diseases such as cholera, typhoid fever, and dysentery can be passed on like wildfire through fecal contamination of water supplies. It is for this reason that modern internal plumbing, efficient sewers, water chlorination, and other measures have done so much to cut down the incidence of epidemics. Finally, feces contain the bile pigments released by the liver, and this gives them their color.

The act of eliminating the feces is *defecation.* This can be taken care of by the natural peristaltic action of the rectum, and can be aided by compression of the diaphragm and of the muscles of the abdominal wall. In infants, the process takes place whenever the rectum is full enough to stimulate peristalsis. It is with some trouble that we "toilet-train" infants and teach them to control the act at least until they can reach the bathroom.

Defecation is usually accompanied or preceded by the escape of intestinal gases. This is *flatus* (flay'tus; "blowing" L). The gas is harmless but contains small quantities of volatile compounds formed by bacterial action in the large intestine and these have an unpleasant fecal odor.

Whenever the interval between defecations is greater than normal for a particular individual, he is suffering from *constipation* ("to press together" L). This happens when the peristaltic action of the large intestine is slower or weaker than usual. The slower passage of food through it allows opportunity for a greater subtraction of water. The feces become harder than usual, are compacted and pressed together, and are eliminated only with great difficulty. Since peristalsis is stimulated by the presence of roughage in the feces, cereals with high content of bran will sometimes counteract constipation. Certain chemicals, found naturally (as in prune juice) or formed synthetically and incorporated into proprietary "laxatives," may also be used. The walls of the large intestine can be lubricated so as to make it easier for the hardened feces to pass along, and it is for this reason that mineral oil and castor oil are effective. Epsom salts brings about an influx of water from the body into the intestines, softening the feces; and, finally, there can be the direct washing action of warm water (an "enema") introduced through the anus.

In reverse, it may also happen that the large intestine pushes food through too quickly, sometimes because of infections that powerfully irritate the intestine. In such a case, little fluid has had a chance to be absorbed and the feces are watery. This is *diarrhea* ("a flow through" G). Diarrhea, like vomiting, can be very debilitating, even quite apart from any infectious disease that may be giving rise to it. It represents a loss of needed water and inorganic ions. In the case of infants, who have a very small water reserve, unchecked diarrhea can easily be fatal.

9

OUR KIDNEYS

CARBON DIOXIDE AND WATER

As I have explained in the past few chapters, oxygen is brought into the blood by the respiratory system and food is brought in by the digestive system. Both are carried to the individual cells of the body by the circulatory system. In the cells, the food and oxygen are combined to produce energy. In the process, however, the substance of the food and oxygen is not destroyed. As far as the body is concerned, the atoms making up the molecules of oxygen and of the various components of food are everlasting. They are merely rearranged into new combinations.

The molecules of carbohydrates and fats are made up of atoms of carbon, hydrogen, and oxygen. When those molecules are combined with additional oxygen atoms (*oxidation*) the products are carbon dioxide (made up of atoms of oxygen and carbon) and water (made up of atoms of oxygen and hydrogen). Protein molecules are far more complicated in structure. They contain not only atoms of carbon, hydrogen, and oxygen, but also numerous nitrogen atoms, plus a scattering of atoms of sulfur, phosphorus, iron and so on. Proteins, therefore, on combination with oxygen give rise not only to carbon dioxide and water but also to nitrogen-containing compounds and to substances containing the other atoms mentioned.

All these products of oxidation may be looked upon as wastes analogous, in a way, to the ashes left behind by a fire that has burned all it can. The process of getting rid of wastes is excretion ("to separate out" G) and the organs primarily involved in the excretion of wastes make up the *excretory system* (eks'kreh-toh'ree).

Carbon dioxide is a gas, and in all animals is handled very much as oxygen is, except that it moves in the opposite direction. In animals simple enough to subsist on the direct diffusion of oxygen from the oxygen-rich environment into the oxygen-poor cells, there is a contrary diffusion of carbon dioxide from the interior of the cell, rich in the gas, to the outer environment, poor in it.

More complicated creatures with specialized organs for oxygen absorption and a circulatory system for oxygen transport make use of those same devices for carbondioxide excretion. Thus, as cells combine food and oxygen, the carbon dioxide that is produced diffuses out of the cell first into the interstitial fluid and eventually the blood. In a way, the blood can handle carbon dioxide more easily than it can handle oxygen, for carbon dioxide is the more soluble by far. Whereas 100 cubic centimeters of water at body temperature will dissolve only 2.5 cubic centimeters of oxygen, it will dissolve fully 53 cubic centimeters of carbon dioxide. In addition, some of the carbon dioxide can tie loosely onto portions of the hemoglobin molecule that are not involved in oxygen transport.

Partly dissolved in water, partly combined with water to form carbonic acid, partly combined with hemoglobin, the carbon dioxide finds its way at last into the capillaries lining the alveoli of the lungs. There, while oxygen passes from the alveoli into the bloodstream, carbon dioxide passes from the bloodstream into the alveoli. Whereas inspired air is only 0.03 per cent carbon dioxide, expired air is about 5 per cent carbon dioxide.

Water excretion is even less of a problem. In fact, it is no problem at all, since water produced by the oxidation of foods joins the water that already makes up 60 per cent of the human body and is not to be distinguished from it. Actually, water should not even be considered a waste, because it is an absolutely essential component of living tissue, and for any creature living on dry land the problem is not to get rid of it but to conserve it. The body unfortunately cannot help losing water for a variety of reasons. In the first place, the alveoli are permeable to water and are always moist. They must be moist because the diffusion of oxygen and carbon dioxide can only take place after the gases have dissolved in the film of water lining the alveoli. A dry alveolus would not function. Expired air is therefore always saturated with water vapor, and, except on the rare occasions when the local atmosphere itself happens to be saturated with water vapor, this means that our bodies lose water with every breath. In addition, we maintain our constant body temperature, despite changes in the temperature of the environment, very largely through the agency of perspiration. This is an efficient air-conditioning system but it can be prodigal in its use of water, which is then lost to us. In the last place, we need water as a solvent for the wastes arising from protein, so some water is unavoidably lost in the process of getting rid of those wastes.

There are some animals that have developed ways of cutting down these water losses to the point where the water formed by the oxidation of food is sufficient to replace what water is lost. Such animals (usually adapted for desert life) never have to drink but can make out perfectly well on food alone—because, first of all, food is never really dry. Vegetation is often 80 to 90 per cent water, and fresh meat is 70 per cent water. For that matter, bread is at least 30 per cent water if it is fresh, and even something as dry as dried lima beans is more than 10 per cent water. Add to this the water arising from the oxidation of foodstuffs and it is not so surprising that some water-economizing desert animals never have to drink.

The human body, on the contrary, cannot conserve water well enough for food alone to serve as an adequate renewal. The average adult loses up to two liters a day through his lungs and skin and in his urine. (As a result of vomiting or diarrhea, or through excessive perspiration on hot days or during unusually hard work, he can lose considerably more.) For that reason it is necessary that an adult drink some two liters of water (or watery fluids such as fruit juice, beer, etc.) each day. Ordinarily, there is no difficulty about this (provided water is available): when water loss has reached 1 per cent of the body weight the sensation of thirst is experienced, and a thirsty person needs no urging to drink.

The direct stimulus of thirst seems to arise when the pharynx dries and salivation is partially suppressed because of water shortage. However, the deeper cause is the increased concentration of dissolved substances in the blood because of water shortage. Thus, mere wetting of the mouth and throat does not relieve the symptoms of thirst for more than a moment. On the other hand, the introduction of water directly into the stomach does relieve them even though the mouth is not directly wetted.

Thirst is more uncomfortable, more demanding, and less easily withstood than is hunger. This is understandable, since while the average well-nourished man has a considerable food store to draw on in case of emergency, the water store is far smaller. If water is not available, the individual approaches a state of collapse when water loss passes 5 per cent of the body weight and is near death when water loss passes 10 per cent of the body weight. This may seem to compare favorably with the amount of weight of fat lost during starvation, but water loss proceeds far more quickly. The limit of human endurance of thirst is reached in a matter of days, whereas endurance of hunger can proceed for weeks. Water reaches the intestines rapidly once it is swallowed and is there rapidly absorbed, diluting the overconcentrated blood. Thirst therefore vanishes almost with the drinking.

What about the excretion of wastes other than carbon dioxide and water produced by the combination of proteins with oxygen? How does an organism get rid of the waste nitrogen atoms which, next to carbon, hydrogen, and oxygen, are the most common in the protein molecules? It might seem that the logical way would be to allow it to form nitrogen gas and, like carbon dioxide, have it eliminated at the lungs. Alas for logic! The production of gaseous nitrogen is an energy-consuming process of a kind of which no organism above the level of certain bacteria is capable. Then, even if nitrogen were formed, it is even less soluble in water than is oxygen, and its transport, in quantity, by the bloodstream would pose a pretty problem.

An alternative is to produce *ammonia* as a product of the combination of protein with oxygen. Ammonia, like nitrogen, is a gas (with a molecule containing nitrogen and hydrogen atoms) and can be formed by processes that are not energy-consuming. Furthermore, it is extremely soluble in water and its transport by the bloodstream would involve no problem. As a matter of fact, many sea organisms do indeed excrete nitrogen in the form of ammonia.

There is a catch to this far as we ourselves are concerned, and it is a serious one. Ammonia is extremely toxic to all forms of life. A thousandth of a milligram of ammonia in each liter of blood would be enough to kill a man. The sea creatures that excrete ammonia can get away with it because they have a vast ocean into which to dump the gas as quickly as it is formed. In the ocean, the ammonia forms a solution far less concentrated than even the tiny value that is fatal to life. Nor will the ocean become more concentrated with time, for there exist microorganisms within it that make use of the ammonia, combining it with other compounds and rebuilding protein out of it.

Land creatures, with a limited supply of body water at their disposal, cannot make use of ammonia as a waste. Many organisms use instead an easily soluble solid substance called *urea* (yoo-ree'uh). The molecule of urea is built up of the fragments of two ammonia molecules and a carbon dioxide molecule. Using it as a nitrogen waste is slightly more inefficient by a couple of per cent than using ammonia would be because, for one thing, urea formation is an energy-consuming process. However, urea is far less toxic than ammonia and that more than makes up for any slight loss in efficiency.

Urea can be allowed to build up to a fair concentration. A hundred milliliters of blood will contain up to 33 milligrams of urea, which is a hundred thousand times and more the quantity of ammonia that would be fatal. Consequently, it would take a hundred-thousandth of the quantity of water to excrete the day's supply of urea that would be required to excrete the day's supply of ammonia. Quite a saving of water.

The change is conspicuous in amphibians, which spend their early life as water creatures and their later life as land creatures. The tadpole has gills and tail, then loses both and develops lungs and legs instead. This change is noticeable and startling. Hidden from our eyes is another change just as important, one without which the other changes would be meaningless as far as survival is concerned. For whereas the tadpole excretes ammonia, the adult frog excretes urea.

Reptiles and birds are plagued by an even sharper water shortage than are amphibians. Amphibians lay their eggs in water, but reptiles and birds lay eggs on dry land. The water supply within the egg, at the disposal of the developing young, is particularly limited and even urea will not serve as a means of excreting nitrogen. Although urea is relatively nontoxic, it is not absolutely nontoxic; it will kill if the concentration rises high enough. Reptiles and birds therefore excrete nitrogen in the form of *uric acid*. This is a compound with a relatively complicated molecule built up of fragments of four ammonia and three carbon dioxide molecules (plus several additional atoms). Uric acid is quite insoluble, so that in emergencies, as in the egg, it can be tucked away in odd corners of the organism without tying up any significant amount of water.

For mammals, the water shortage eases up slightly. The developing young, for which, in birds and reptiles, the water shortage was most extreme, remains in mammals amid the watery tissues of the mother. For that reason, urea will do as the excretory form of nitrogen, and the human being, like other mammals, excretes urea.

THE EXCRETORY SYSTEM

Urea cannot remain in the blood, of course. It must somehow be brought to the outside world and there discarded. In many nonchordates and in some primitive chordates too, this is done by means of individual microscopic tubes into which there filters water from the plasma. The wastes accompany the water, are led through the tubes to the surface of the body, and are discharged into the watery environment outside. In the vertebrates the number of

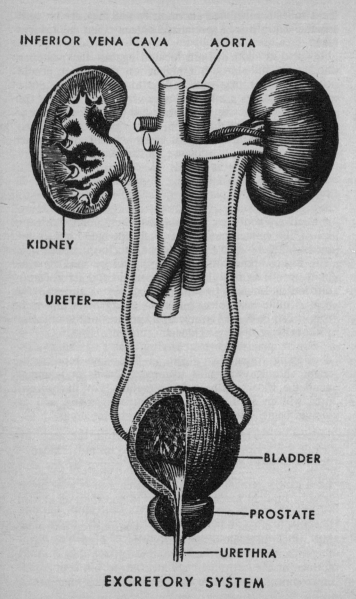

INFERIOR VENA CAVA   AORTA

KIDNEY

URETER

BLADDER

PROSTATE

URETHRA

EXCRETORY SYSTEM

these tubes is multiplied enormously and they are brought together into a pair of specialized organs called the *kidneys* (AS).

In man, the kidneys are located against the posterior wall of the abdomen, but higher up than most people suspect. The average man, asked to point to the place where his kidneys are located may very likely indicate the small of his back. Actually, they are just below the diaphragm, in front of the bottom-most ribs and behind the liver and stomach. The right kidney, which is crowded above by the liver, is usually a bit lower than the left. Individually, the human kidney is a liver-colored organ, about 4 to 5 inches long, 2 to 3 inches wide, and 1 to 2 inches thick. It weighs about half a pound and has the familiar lima-bean shape. In fact, there is a bean called the "kidney bean" because of its shape and color. The kidneys lie outside the peritoneum but are firmly held in place by connective tissue and a cushion of fat. They consist of an outer *cortex* ("bark" L, a word applied by analogy with the bark of a tree to the outer portion of other objects) and an inner *medulla* (me-dul′uh; "marrow" L, applied, by analogy with bone, to the inner portion of other objects).

The kidney consists largely of a huge mass of filter tubes or *nephrons* (nef′ronz; "kidney" G). These are also called *uriniferous tubules* ("urine-carrying" L). There are roughly a million nephrons in each kidney and these represent considerably more than we absolutely need. It is possible for a man to endure the loss of many nephrons through disease, or even the complete removal of one kidney, and still lead a normal life.

Blood is led to the kidneys directly from the aorta by way of the short, thick *renal arteries* (ree′nul; "kidney" L). The importance of the kidney is shown by the fact that at any given moment, as much as one quarter of the blood supply may be passing through them; as much through the single pound of the two kidneys as through the nearly hundred pounds of muscle in the body. The renal artery breaks up into numerous arterioles, each of which, along its path, breaks up further into a mass of coiled and intertwined capillaries shaped in a tiny ball. These combine again into the arteriole, which then passes on to divide into capillaries in the usual manner. (These final capillaries feed the kidney tissue.)

The section of the arteriole before this coiled ball of capillaries is the *afferent arteriole* (af'ur-ent; "carrying toward" L), and the section after it is the *efferent arteriole* (ef'ur-ent; "carrying away" L). The coiled ball of capillaries itself is called the *glomerulus* (gloh-mer'yoo-lus; "small ball of wool" L, which it rather resembles). The blood is brought back from the kidney via the *renal vein* to the inferior vena cava.

The blood passing from the arteriole to the glomerulus suddenly finds the total cross-sectional area of the vessels greatly increased and blood flow is consequently slowed. There is ample time for water, ions, and small dissolved molecules such as urea to diffuse outward from the glomerulus into a section of the nephron which encloses the ball of capillaries like a clutching hand. This enclosing section is called *Bowman's capsule* after Sir William Bowman, the British surgeon who first described it.

It is absolutely essential that the blood be continuously filtered in this fashion; so essential, in fact, that the kidneys have their own special device for maintaining blood flow through themselves at the proper rate. If, for any reason, blood pressure falls so that the blood flow through the kidneys is decreased below the normal range, the kidney is stimulated to produce a substance called *renin* (ree'nin), which it discharges into the blood. This in turn stimulates the contraction of arterioles, reducing the volume of the circulatory system and increasing the blood pressure to a safe level. Where the blood flow through the kidneys is interfered with for reasons other than low blood pressure—as by an abnormal thickening of the walls of the renal artery—the kidneys will produce a more or less permanent hypertension to make up for it. (However, most cases of hypertension are not caused by the kidneys, but are of unknown cause.)

The hard-working kidney is sometimes subjected to inflammation through bacterial infection, or, occasionally, for other reasons. This is called *nephritis* (nef-ry'tis; "kidney inflammation" G). Where kidney tissue undergoes degeneration or destruction without inflammation, the disease is *nephrosis* (nef-roh'sis). In view of the importance of the kidney, both diseases can be extremely serious. Both are sometimes lumped under the name of *Bright's disease,* after the British pathologist Richard Bright, who first described their symptoms systematically.

The fluid that diffuses from the glomerulus into Bowman's capsule has left the body at that point, and is carried through a series of tubes eventually eliminating it into the outer environment altogether. The kidney's work is not done, however; in fact, the truly important portion has not yet begun. What gets into Bowman's capsule is an almost indiscriminate filtrate of the plasma. It contains everything except some of the protein molecules that are just too large to get through. It contains not only urea, which it is desirable to discard, but also a great deal of water, plus useful ions, plus glucose, plus numerous other substances that it is not desirable to discard.

Bowman's capsule leads into a *convoluted tubule* (that is, a coiled little tube) where the desirable material is reabsorbed. By the time the fluid has passed through this

TUBULE STRUCTURE OF KIDNEY

tubule it has become a relatively concentrated solution, carrying wastes only. In a creature dwelling in fresh water and therefore having no fear of water shortage, as in the case of the frog, the convoluted tubule is comparatively short, and reabsorption of water is only moderate. The fluid eliminated is very dilute. In a land animal such as man, reabsorption must be more extensive, since water cannot be wasted. In man, therefore, the convoluted tubule is divided into two parts: the part leading from Bowman's capsule is the *proximal convoluted tubule* and the more distant portion is the *distal convoluted tubule*. Between the two parts and connecting them is a long, straight, and particularly narrow section that is bent upon itself like a hairpin. This is *Henle's loop* (hen′liz), named for Friedrich Henle, the German anatomist who first described it.

This additional length of the tubule increases the efficiency of reabsorption, and makes its extent adjustable. In man, about 80 per cent of the water and ions that pass out of Bowman's capsule are reabsorbed in the proximal convoluted tubule. This represents a minimal reabsorption, and if a person has been drinking water copiously, little more is absorbed in Henle's loop. The fluid finally excreted is fairly dilute. Under ordinary circumstances, however, considerably more is absorbed in Henle's loop. The greater the dehydration of the individual, the more (up to a certain maximum) is reabsorbed.

On the whole, 120 cubic centimeters of fluid filters out of the glomerulus every minute. This amounts to 50 gallons a day, but 99 per cent of this is reabsorbed through the convoluted tubule and Henle's loop. This ability to reabsorb water is regulated by a hormone formed by the *pituitary gland* (pi-tyoo′i-ter′ee), a small organ at the base of the brain. In some individuals the supply of this hormone dwindles, and reabsorptive ability likewise dwindles. The fluid produced by the kidney (*urine*) is therefore both copious and dilute.

A disease characterized by an abnormal quantity of urine, or *polyuria* (pol-ee-yoo′ree-uh; "much urine" G), is called *diabetes* (dy-uh-bee′teez; "siphon" G), because water seems to pour through the body, in at one end, out the other, as through a siphon. This particular variety of the disease is *diabetes insipidus* (in-sip'i-dus; "tasteless" L), since the very dilute and watery urine is virtually tasteless as compared with another form of diabetes where it is sweetish. A person with diabetes insipidus must necessarily replace the lost water and is therefore plagued by incessant thirst.

The kidney may also fail in its ability to reabsorb water through Bright's disease. This alone may be compensated for by frequent drinking, as in diabetes insipidus. Eventually, however, failure of kidney function can progress to the point where urea is not efficiently filtered out of the bloodstream in the first place. The urea concentration in blood rises, a condition called *uremia* (yoo-ree'mee-uh; "urine in the blood" G), and death ensues.

But to return to the tubules. Finally, having passed through the two sections of the convoluted tubule and Henle's loop between, the fluid enters a *collecting tubule,* which is, indeed, just a duct for waste matter and receives the influx of numerous convoluted tubules. The fluid is now properly to be considered urine (from a Greek word of uncertain derivation). The compounds urea and uric acid were so named because they were first discovered in urine. The individual tubules of the kidney are microscopic but not short, for if they were straightened out they would be an inch long or more. All the tubules in both kidneys would come to a length of about 40 miles. And although each collecting tubule receives the merest trifle of urine—it would take two years for an individual nephron to deliver a single cubic centimeter of urine—all of them, working together, deliver something like a cubic centimeter of urine each minute.

The tubules collect into larger ducts until finally all merge into the *renal pelvis,* a space that fills most of the interior of the concave side of the kidney. The renal pelvis narrows into a tube called the *ureter* (yoo-ree'ter; "urinate" G), which leads down along the rear wall of the abdomen some 10 to 12 inches. Just in front of the lower reaches of the intestines, the two ureters, one from each kidney, open into a muscular-walled sac, the *urinary bladder.* (This is often referred to simply as the "bladder," but there are other bladders in the body—notably the gall bladder.) The bladder serves as a storage place for urine, so that although the kidneys constantly form the fluid we need not constantly eliminate it, but can do so at intervals, according to convenience. The muscles of the bladder slowly relax as urine enters until it has expanded upward into a sphere bulging into the abdomen. At maximum expansion it can hold more than a pint of urine.

The ureters enter the bladder at points near the bottom. At the very bottom of the bladder is a thicker tube, the *urethra* (yoo-ree'thruh), a word which is but another form of "ureter." Through the urethra the urine is finally carried to the outer world. The urethra is considerably different in length in the two sexes. In women it is not more than 1 to 1½ inches long. In men, however, it is some 8 inches long, passing through the length of the penis. About the male urethra is the prostate gland which I shall talk about later in the book.

The exit way from the bladder to the urethra is closed off by a pair of sphincters; because of these, under ordinary circumstance, the urine does not leave the bladder. As the bladder fills up, however, there comes a point where the muscular wall begins to contract rhythmically, setting up an increased fluid pressure against the base of the urethra which makes itself felt by the individual as a desire to urinate. This continues with increasing urgency until urination takes place.

In infants this rhythmic contraction actuates a reflex that relaxes the urethral sphincter and leads to urination at once. As the child grows older, he usually learns (with the more or less forceful encouragement of the parents) to control the reflex. This is most difficult to do during sleep, of course, and bed-wetting may continue for months after a child is otherwise toilet-trained. In some cases, bed-wetting may continue onward into adolescence and even adulthood.

The bladder may be inflamed through bacterial infection, a condition called *cystitis* (sis-ty'tis; "bladder inflammation" G). This is sometimes in association with kidney inflammation, but it is the bladder inflammation that is the more likely to force itself upon the victim's attention, since its most noticeable symptom is a painful urination.

URINE

About one to one and a half liters of urine are formed and discharged per day. It is an amber-colored liquid with a distinct odor not too annoying when fresh. It is free of bacteria to begin with (unless there is an infection of the kidneys or bladder), but if it is allowed to stand in the open, bacteria will infest it and the consequent putrefaction will produce an unpleasant stench.

Thanks to reabsorption of fluid in the tubules, the urine is comparatively concentrated, but at best it is still some 95 per cent water. It is impossible for the human excretory system to reduce the water content any further; and since wastes must be disposed of if life is to continue at all, the urine continues to drain the body's water supply even when a person is in the extremity of thirst. It is for this reason that a human being adrift in a lifeboat cannot help himself by drinking ocean water. The salt content of the ocean must be excreted, and that takes more water than was in the ocean water consumed. The result is a net loss of water and a quicker death. (Sea water contains about 3.5 per cent of salt and other inorganic constituents. Urine contains about 1 per cent of these, so for every milliliter of sea water a human being must pay with three milliliters of urine—a losing proposition.)

The chief solid carried in solution in the urine is, of course, not inorganic, but is the organic compound urea; and it is urea that builds up the concentration of urine to the 5 per cent mark. The actual amount of urea eliminated per day depends on the quantity of protein in the diet for it is from protein that urea is derived. On a good protein-rich diet, over 40 grams of urea can be eliminated each day.

Small amounts of other nitrogen-containing compounds are also eliminated. There is uric acid, for example. We, unlike birds and reptiles, do not form it from proteins. Nevertheless, we do form it from some of the building blocks of the nucleic acids (essential compounds found in all cells). There is also *creatine* (kree'uh-teen; "muscle" G) and *creatinine* (kree-at'i-neen), formed to a small extent from the breakdown of proteins; particularly, as the names imply, from the protein of the muscles. There is even a bit of ammonia formed in the process of urine manufacture. In addition, there are various inorganic ions, the breakdown products of hormones, products formed by liver out of foreign molecules, and so on.

The urine serves not only as a reservoir for wastes but also as a device by which to regulate the concentration of many body components. Any substance, ordinarily useful, which is present in excess, is very likely to find its way out of the body by way of the urine. When that same substance is in short supply, however, urinary loss is cut down, in some cases nearly to zero. The most dramatic example of this arises in those cases where the body's normal supply of a hormone called *insulin** (in'syoo-lin) dwindles. This hormone is necessary for the proper breakdown of glucose by the body. With insulin in short supply, abnormal breakdown products called *ketone bodies,* accumulate, and so does glucose itself.

The ketone bodies are quite dangerous, for in more than a certain minimal concentration they increase the acidity of the blood to the point where the patient goes into a coma and dies. If the disease remains untreated this eventuality is inevitable, but the kidneys postpone the evil day by eliminating all the ketone bodies they can. For the purpose, the volume of urine is increased beyond the normal range, so that the disease is a form of diabetes (but the polyuria is not as extensive in this case as in diabetes insipidus). The individual with an advanced and untreated case of this disease is naturally excessively thirsty and also excessively hungry, because although he eats his body is not using the food efficiently. Even though he may be eating well, he will be losing weight.

The inefficient way in which such a person makes use of his food is most dramatically shown in the fate of his body glucose. The fluid passing out of the glomerulus always contains glucose, of course, but in the course of its passage through the tubule, all the glucose is reabsorbed so that normal urine is glucose-free. In this form of diabetes, however, the concentration of glucose in the blood rises

* This hormone is produced in the pancreas, and it will be considered in greater detail in the companion volume to this book.

to abnormally high levels and it becomes harder and harder for the tubules to reabsorb it all. Finally a point is reached where reabsorption remains incomplete. Glucose concentration is then said to have risen above the *renal threshold* and it makes its appearance in the urine. The presence of a form of sugar in the urine of some people appears to have been discovered in ancient times, when it was noticed that such urine attracted flies. Cautious tasting must have revealed the presence of sweetness, and this particular disease is therefore *diabetes mellitus* (meh-ly′tus; "honey" L). Diabetes mellitus is both more serious and more common than diabetes insipidus. It is the former disease that is usually meant when the word "diabetes" is used by itself.

The inefficiency involved in urinating away good food is not entirely without its compensations. If the glucose concentration in the blood were allowed to increase without check, the viscosity of the blood would increase to the point where circulation would be fatally interfered with. Excretion of glucose is wasteful but it prolongs life.

From the diagnostic standpoint, the value of urinary glucose is that its presence can be easily tested and is a sure indication of well-advanced diabetes. The disease can now be easily treated by injections of insulin obtained from domestic animals slaughtered for food, and it pays to diagnose diabetes at the earliest possible moment, *before* glucose appears in the urine. This is done by tests involving the blood itself.

The presence of other abnormal constituents in the urine may also be indicative of disorders in body chemistry; disorders, fortunately, which are not usually as serious as diabetes. Sometimes amino acids, the building blocks of proteins, appear in the urine in more than normal quantity; sometimes certain breakdown products appear. There is, for instance, a compound called *homogentisic acid* (hoh′-moh-jen-ti′sik) that appears in the urine of some people who from birth are lacking in the ability to break down an amino acid called *tyrosine* (ty′roh-seen) in the proper manner. The urine containing homogentisic acid will, under the proper conditions, turn black upon standing, but despite this startling fact, the disease is not in the least serious.

The principle of kidney action rests, of course, upon the fact that all the wastes to be eliminated can be flushed out of the body by a current of water. One might assume then that those wastes are soluble in water. Unfortunately this is not exactly so. Some of the substances eliminated by way of the kidneys are not at all soluble in water. As an example, there is uric acid. Although the human being eliminates this in small quantities in the urine, it is quite insoluble. Other mammals break up the uric acid into a more soluble compound, but primates, including man, lack the ability to do this. Again, some of the inorganic ions ordinarily present in urine can combine to form insoluble substances such as calcium phosphate and calcium oxalate.

The question then is this: How are these insoluble materials disposed of? The answer is that even solids can be carried along by a current of water, if present in small enough pieces. Urine often contains microscopic crystals of solid matter that are carried along without particular difficulty. These crystals do not have much of a tendency to aggregate under ordinary circumstances. The reasons for this are not entirely established, but one reasonable guess is that the individual tiny crystal is coated with a thin layer of some protective substance, such as protein or mucopolysaccharide, that keeps them from aggregating even when they make contact. In some individuals, this protective device fails and there is then a tendency for the crystals to merge into *kidney stones,* or *urinary calculi* (kal'kyoo-leye; "small stones" L). These may easily become too large to pass through the ureters. In some cases, calcium phosphate stones (which grow quickly) may virtually fill the renal pelvis. Calcium oxalate stones, which grow more slowly, are jagged and irregular and cause intense pain (like a hugely magnified and unending stomach ache) when trapped in the ureter. The pain of kidney stones is sometimes called *renal colic* because of this resemblance to intestinal pains, though it has nothing to do with the colon.

Organic substances will also, though more rarely, form stones. The amino acid *cystine* (sis'teen) is a normal component of proteins and is the least soluble of the amino acids. It is sometimes excreted in small quantities in the urine and can collect to form a stone in the bladder. In fact, cystine was first isolated from such a *bladder stone,* and its name comes from the Greek word for "bladder."

Uric acid will also form stones, and here a new area of danger arises. Sometimes uric acid is deposited in the joints of the extremities, particularly of the big toe, to give rise to the extraordinarily painful disease *gout.* (This word arises from a Latin word meaning "drop" because in the Middle Ages there arose the misconception that gout was caused by the gathering of some fluid in the joints, "drop by drop.") Gout seems to have been more prevalent in previous centuries than now, partly because conditions which were once diagnosed as "gout" are now diagnosed as some form of arthritis.

10

OUR SKIN

SCALES AND EPIDERMIS

In primitive animals, both unicellular and multicellular, the outer surface makes contact with the surrounding environment, and it is at the outer surface that most of the interactions with the environment take place. As animals grew complex, however, more and more of these interactions were performed on interior surfaces. The food canal was brought into being and placed in the interior. The respiratory and excretory systems evolved interiorly. Only very small portions of the outer surface came to be involved with the intake of food and air and with the outlet of wastes. The outer surface, except for these minor regions, could therefore be preserved for the passive task of protection.

Many phyla of animals developed shells of one sort or another to serve as such protection. These, nonetheless, added weight and reduced the sensitivity and responsiveness of the creature to stimuli from the outer world, and their mobility as well. The chordates, with their internal bracing, could afford to be unshelled and the risk that was therein involved was more than made up for by the improvement in efficiency. Even so, the victory of the unshelled was not a quick one. Among the nonvertebrate chordates the tunicates, at least, developed a tunic serving the function of a shell. As for the vertebrates, the first two classes backed up their internal skeleton with an external skeleton as well. In fact, as I explained in Chapter 1, bone was first developed not as an internal skeleton (which remained cartilage for millions of years) but as external armor. Even in man the collarbone and the bones of the skull are inherited remnants of this external armor, drawn inward now beneath the skin.

The armored vertebrates of the sea lost out to the more efficient sharks and bony fish, which abandoned shells and relied instead on the speed and maneuverability made possible by the reduction in mass. (Nevertheless, even in classes developed later there continued the tendency to fall back upon the defensive safety of armor. Among creatures now living, the turtles among the reptiles and the armadillos among the mammals are examples. It never proved a spectacularly successful maneuver, and yet turtles and armadillos still survive, so we can't toss it aside as altogether unsuccessful, either.)

The loss of a bony armor did not mean that fish were entirely naked and unprotected. In the place of bone they developed light, cleverly overlapping scales—tough and flexible. Among the land vertebrates, scales of a somewhat different variety developed, a more superficial type, easily capable of being shed and replaced from below. These are best developed among the reptiles and the ability of the snake to shed its scaly cover periodically is well known. The reptilian scale persisted in specialized fashion among the warm-blooded descendants of the reptiles, the birds and the mammals. They are to be found, for instance, on birds' legs (look at a chicken leg next time you have a chance) and on rats' tails. Even in a human being, fingernails and toenails are a version of the reptilian scale.

However, birds and mammals are warm-blooded and, in order to be effectively so, insulation is required against the excessive loss of heat to the outer world. Scales are not sufficiently insulating unless they can be made loose in order that they might trap a layer of unmoving air next to the body. (Unmoving air is an excellent insulator.) In birds such loosened scales developed into feathers, and in mammals they developed into hair.

Feathers are the more efficient of the two as insulators. They have other essential uses as well. The large feathers of the wings make flight possible and the large feathers of the tails serve as balancing devices. The connection between feathers and flight seems to result in the fact that no flying bird is without a full complement of feathers even in warm environments (with a few minor exceptions, such as the unfeathered heads of vultures). Even flightless birds keep their feathers, although these can become pretty ragged in some cases, with little left beside the central shaft. Hair, on the other hand, has no essential purpose other than that of insulating against heat loss (though some animals have developed specialized uses), so hair-poor animals are not uncommon in the tropics. Such creatures as the elephant and hippopotamus possess few hairs. Whales, who use blubber as insulation, are completely hairless, although a few bristles make their appearance in the embryo. Beneath the scales, feathers, or hair is the soft and sensitive skin of the vertebrate which still serves as protection. Through its unbroken surface, microorganisms and foreign bodies cannot enter; and it can withstand the buffeting of rain, wind, heat, and cold as internal organs could not.

The characteristic protein of skin and skin appendages is *keratin* (ker'uh-tin; "horn" G, because it occurs in horns). Keratin is an extraordinarily tough protein, insoluble, indigestible, and relatively immune to damage by shifting environment. The toughness of the protein component is reflected in the skin itself.

Skin is divided into two main regions. Inside, under the part we actually see, is the *dermis* ("skin" G). This is a living tissue, rich in nerve endings, blood vessels, and various glands. Beneath it is a layer of connective tissue containing the subcutaneous fat I mentioned in Chapter 8. Above the dermis is the portion of the skin we actually see, the portion that fronts the outside world. This is the *epidermis* ("upon the skin" G) and it is dead. The cells at the base of the epidermis are alive, and are constantly growing and multiplying so that cell after cell is pushed upward and away from the blood supply of the dermis. Without a blood supply, the cell dies and much of it, aside from the inert keratin, atrophies. The vicissitudes of existence are constantly rubbing away some of this dead material from the surface of our body, but this is constantly being replaced from below, and we retain our epidermis ever fresh.

This process takes place rather quickly, too. It has been shown that the epidermis of a rat's foot undergoes complete replacement in three weeks and those regions of man's epidermis most exposed to friction may be equally renewable. The fact that epidermis is constantly growing means that it is a region we can reconstruct, or *regenerate,* if a section of it is destroyed. The dermis itself is not so easily reconstructed after destruction. The breach is indeed healed over, but only by a bridge of connective tissue. The specialized structure of that section of the dermis is lost and the featureless section of connective tissue replacement forms a *scar* ("fireplace" G, because burns are a common cause of scarring).

Reptilian scales, avian feathers, mammalian hair are all epidermal in origin, and like the epidermis itself are constantly being shed and replaced. The scales of fish are of dermal origin and their loss is a more serious matter.

The surface of the dermis is ridged and possesses tongue-like processes. The epidermis, for the most part, fills in the spaces between the processes and covers it all smoothly. On the palms of the hands and the soles of the feet, however, the epidermis rises and falls with the processes, so small parallel lines extending in gentle curves are to be found there. On the balls of the most distal joints of the digits the lines fall into whorls and loops. The purpose of these tiny ridges is to supply a surface of greater friction in order that feet grip the ground better in walking and hands grip anything better in grasping. They serve the purpose that tire treads do. The exact pattern of the ridges upon the palms and soles is highly individual; if two "fingerprints" are found to be identical in every respect, it is quite safe to assume that the same person made them both. (The prints are outlined by tiny beadlets of sweat and oil produced by glands in which the palms are particularly rich. The film of moisture produced in this way further improves the gripping capability of the hands and feet.)

The soft sensitivity of the skin is not to be attributed to the epidermis, which is itself dead and insensitive. The epidermis is thin enough so that the nerve endings in the dermis are sufficiently close to the surface to supply the sensitivity. Where areas of the skin are chronically subjected to friction, in response the epidermis thickens, to accentuate the protection it offers. A *callus* ("hard skin" L) is formed. Thus the soles of the feet are commonly callused among those who habitually walk barefoot, and the palms of the hands are callused in laborers. (In the days when most labor was manual labor, soft hands were the sign of aristocracy, and one of the feats of Sherlock Holmes was the ability to discern the specialized occupation of someone from the type of calluses he possessed. It was because the bluish veins were visible through the soft, uncallused, and un-weatherbeaten skin of the hands and arms of those who did not labor, that aristocrats were said to be "blue-blooded.")

The deadness of the epidermis is quite apparent on callused spots, for there the skin is notably hard and inflexible, as well as relatively insensitive. It is as though the living and sensitive dermis has exchanged its ordinary thin latex gloves for a pair made of leather. Sometimes an area of excessive irritation or pressure, such as that caused by ill-fitting shoes, may produce an area of abnormal epidermal thickening on a toe. This is a *corn,* and it can become quite painful.

A word derived from the same root as "corn" is *horn,* and indeed the two are quite similar in chemical structure. The horns and antlers of various animals are keratinized and hardened modifications of the epidermis. So are the hoofs of various grazing animals and the claws of the various carnivores. We ourselves have such horny outgrowths in our fingernails and toenails, which are analogous to claws and hoofs. Our fingernails are no longer very useful to us as weapons of offense or defense (although women have been known to use them with fearful effect). Nevertheless, they stiffen the tips of the fingers and, if allowed to grow, offer thin, hard surfaces which can be used for such delicate tasks as picking up pins or prying into narrow crevices.

Skin protects not only against the mechanical shocks, the blows and scrapes, of the environment; it protects also against the impingement of various forms of energy. Before mankind's technology had developed to the point where it could create concentrations of types of energy with which the body was not designed to cope, the chief form of concentrated energy encountered was sunlight. Most animals are protected from sunlight by a thickness of water (if they are sea creatures) or by an intervening layer of dead matter (if they are land creatures). Scales, hair, and feathers effectively absorb the energetic rays of the sun without harm to themselves, and even the frog, with none of these skin appendages, has at least a thick coating of mucus.

Man is unusual in that his dry naked skin is exposed to the sun with only the relatively thin layer of epidermis as protection. To the ultraviolet rays of the sun, the epidermis of those men with fair skins is quite transparent and might as well not be there.

Ultraviolet light is energetic enough to bring about chemical changes within the cell. Some of these are beneficial. For instance, the dermis contains a type of *sterol* (stee'role; "solid alcohol" G, which is an adequate description of its chemical nature) that in itself is of little value to the body; but under the influence of ultraviolet light it undergoes a slight change that converts it into a form of vitamin D. (This is why vitamin D is called the "sunshine vitamin" by advertising men. It is not in sunshine, but it can be produced in the skin by sunshine.)

Vitamin D is essential to the proper formation of bone (see Chapter 3), and since it is present in very few articles of diet, there was, before the 20th century, constant danger of improper bone formation among children born at the start of the northern winter. The sun was almost the only agency by which the vitamin could be obtained. Considering that man originally evolved as a tropical animal, you can see that this reliance on the sun was safe, to begin with.

When man migrated to the north, however, he reached regions where the sun was in the heavens for but a few hours a day for much of the year (and low in the heavens at that, so most of its ultraviolet was absorbed by the atmosphere). The vitamin was not formed in sufficient quantity, and rickets was the result. When vitamins were discovered and the cause of rickets worked out, fish-liver oils (particularly cod-liver oil), which were rich in vitamin

D, became a favored beverage for the young. Modern vitamin preparations are just as effective, and fortunately far less fishy in aroma. In addition, food such as milk and bread can be irradiated (that is, exposed to ultraviolet light) so that some of the sterols they contain may be changed to compounds with vitamin D activity.

However, this example of vitamin D formation as a beneficent result of exposure to ultraviolet is rather the exception. Other chemical reactions brought on by the energetic action of ultraviolet light are harmful and the skin can respond by an inflammation called *sunburn*. The condition is in every respect a burn and, as those who have suffered it know, it can be uncomfortable and painful.

More serious still is the fact that the ultraviolet in the sun, like energetic radiation generally, can induce cancer. The ultraviolet light is by no means as dangerous as the still more energetic radiations of the X-ray machine and of radioactive substances, but constant exposure to sunlight does increase somewhat the chances of contracting skin cancer. As protection against the irritating effects of ultraviolet light, human skin is equipped with the capacity to form a dark brown pigment called *melanin* (mel'uh-nin; "black" G). This can absorb ultraviolet light without harm to itself and thus acts as a protective umbrella over the regions beneath. It is in tropical areas, where the sun is strongest, that the possession of considerable melanin in the skin is most valuable and it is there that evolutionary processes have slowly increased the quantity of pigment from generation to generation. It is melanin in quantity that is responsible, then, for the dark rich color of tropical peoples such as the Negro of Africa, the Dravidian of India, the Aborigines of Australia, the Papuans of Melanesia, and the Indians of tropical America. Even among Europeans there is a tendency for increasing swarthiness of skin as one proceeds southward.

The pale skins to be found in northern Europe may, on the other hand, also be brought about by evolutionary pressures. Where the sunlight is weak, the presence of melanin is unimportant. Instead, it is better to keep the epidermis transparent so that as much of the weak sunlight as possible reaches the dermis and produces the necessary vitamin D. Under conditions of low melanin content the skin is pale or "white" but permits the redness of the blood in the blood vessels of the dermis to show dimly through so that the actual color is what we call "flesh color."

The formation of melanin is stimulated by exposure to sunlight. This is most apparent among people who are intermediate in melanin content, those who possess little enough to be distinctly "white" in color but possess enough to be "brunet" in complexion. Exposure to the sun darkens the skin to produce a "tan."

When people are very fair they may lack not only melanin but even the capacity to form much. Since hair and eyes also owe their color to melanin, such particularly fair people are likely to have blond hair and blue eyes, and it is these who most often have the characteristic of burning rather than tanning. Some light-skinned individuals produce melanin in localized spots when exposed to sunlight. This is particularly true among those whose light hair contains a reddish pigment that would ordinarily be drowned out if much melanin were also present. It is these "redheads" that form the localized spots of tan we call *freckles* (AS).

Even the lightest normal individuals can form enough melanin to tint their eyes blue. However, individuals are occasionally born who, through a particular flaw in their chemical makeup, are incapable of forming any melanin at all. Skin and hair are white and the eyes are red, because in the absence of pigment the tiny blood vessels are clearly visible in the iris of the eye (the colored part). Such a person is an *albino* (al-by′noh; "white" L). Albinos can occur among any group of human beings; therefore albino Negroes can be found.

Other species of animals may also develop albinism. The white rat and the white rabbit are familiar examples. The white elephant, so venerated in Thailand, is an albino, and one can even have that apparent contradiction in terms, a "white blackbird."

Melanin is not the only skin pigment. A yellow pigment, called *carotene* (kar′oh-teen′) also occurs in the skin. It is a common substance in the plant and animal world (in fact, the name is derived from carrots, which are rich in it) and it is related to vitamin A. Ordinarily it is drowned out by the more deeply colored melanin, but there are groups of people, particularly in eastern Asia, with skins rich in carotene and yet not overrich in melanin, and the result is a definite yellowish tinge in the skin color.

PERSPIRATION

Since the skin is adjacent to the outside world it is important as the means of regulating body heat by offering a radiating surface. The chief source of body heat is not the skin itself, of course, but the internal organs, particularly those engaged in intense chemical activity, such as the liver, kidneys, and heart. The heat produced by those organs is carried off by the blood, which in the course of its circulation distributes the heat evenly. Some of the heat is carried to the dermis and from there part is radiated away. The ease with which this radiation takes place depends on the difference in temperature between the body and its surroundings. When the difference is low, radiation is slow; when the difference is high, radiation is rapid.

In warm weather, when the atmosphere is only a little cooler than the body and the rate of heat loss by radiation is low, the arterioles of the dermis relax, so an unusually large fraction of the blood is in the skin. The slowness of radiation is thus made up for, at least in part, by increasing the supply of heat to be radiated. On the other hand, in cold weather the arterioles of the dermis contract so that the overrapidity of radiation is made up for, at least in part, by reducing the availability of heat for such radiation.

Control by simple radiation is not efficient enough, however, particularly in warm weather when heat must be got rid of quickly. Use is therefore made not merely of simple radiation but of the evaporation of liquid. The conversion of any liquid into its vapor form is an energy-consuming process, and in the case of water the quantity of energy consumed per weight vaporized is higher than for almost any other liquid. The energy for the vaporization is withdrawn from the most convenient place; that is, from whatever the liquid is in contact with. Wet your finger and blow upon it or step directly from a shower into a breeze and the sensation of coolness as vaporization withdraws heat from the skin is unmistakable.

An obvious way of increasing the rate at which water is evaporated from the body is to breathe rapidly and carry quantities of air across the moist surfaces of the mouth, throat, and lungs. We ourselves cannot do this in comfort, but it is the chief method of cooling available to the dog, for instance, which in warm weather will sit with mouth open, quivering tongue extended, and pant.

We don't do this because we have a better device, one the dog lacks. We are equipped with tiny glands distributed all over our skin, about two million of them altogether, the purpose of which is to bring water to the surface of the skin. On the surface this water is vaporized and heat is in this manner withdrawn from the body. The glands are *sweat glands* and the liquid produced is *sweat* (AS), or *perspiration* ("breathe through" L, a reminder of the misconception that the skin breathes through these glands). A sweat gland consists of a tiny coiled tube, the main body of which is situated deep in the dermis. The tube straightens out finally and extends up through the epidermis. The tiny opening on the surface is a *pore* ("passage" G) and is just barely visible to the naked eye.

Sweat is constantly being produced, usually in proportion to the temperature of the environment and, therefore, the need to lose heat by means more effective than simple radiation. In cool, dry weather the amount of sweat produced is relatively small and the rate of evaporation can keep up with it. The skin remains dry to the touch, and you are not aware of sweating. This is *insensible perspiration,* and, for all that you are not aware of it, it can involve the loss of a liter of water per day.

When you are working or playing hard, and heat production of the body is increased, the sweat glands accelerate their production of perspiration. This is also true when the temperature is unusually high. The rate of production may then outstrip the rate of evaporation, particularly if humidity is high, since the rate of evaporation declines with the rise in humidity. Perspiration will then collect on the body in visible drops and we are conscious of sweating. The liquid alone does us little good as far as cooling is concerned; we must wait upon the evaporation. Consequently, when we are visibly sweating we are usually hot and uncomfortable as well, and go around saying, "It's not the heat, it's the humidity"—which is true enough.

On the other hand, when the humidity is quite low, so that the rate of evaporation is high, even hot summer weather does not feel particularly uncomfortable. It is then possible for the temperature of the air to be higher than that of the body, and if radiation alone were involved the body would gain heat; yet, thanks to perspiration and evaporation, the body still feels comfortable. Even when the air is fairly dry, that portion of the air adjacent to our bodies will take up water vapor from our perspiration and will become humid. It is for this reason that it is important there be some sort of ventilation, even if only a small breeze, to replace the humid air adjacent to ourselves with drier air from a distance.

Sweating can be stimulated by emotion or tension as well as heat. This brings on a "cold perspiration," because at lower temperatures the cooling induced by copious perspiration can result in a rather unpleasantly cold sensation. Whereas under the influence of heat the sweat glands of the forehead and neck are most active, under the influence of emotion those of the palms are; and so it is that the palms grow clammy.

Sweat is almost pure water, with dissolved material making up only about half of 1 per cent of the total. Most of that small quantity of dissolved material is sodium chloride, or salt. This loss of salt is ordinarily insignificant, but where perspiration is particularly copious, as much as a liter or a liter and a half of sweat can be lost per hour and then the drain on the body's salt supplies can become appreciable. Loss of water through perspiration naturally stimulates the thirst sensation, and when water is available a copiously sweating person need not be urged to drink. However, such drinking replaces only the water and not the salt. Salt loss, if extreme, can bring about painful cramps, and even where matters are not carried to this point, salt loss will bring on an uncomfortable consciousness of heat. It has become customary, then, for people subjected to intense heat or intense work to take salt tablets with their water.

People adapted to hot, humid climates (Negroes as compared to Europeans, for instance) have more sweat glands and secrete perspiration with a smaller concentration of salt.

As a result of the workings of our sweat-gland air-conditioning system, the temperature of the body is held with amazing constancy between 98 degrees F. and 100 degrees F. The usual "normal temperature" is given as 98.6 degrees F., but this is just an average, and the exact value varies slightly from time to time and from individual to individual. A rise in temperature above the 100-degree F. mark, usually in response to infection, is a *fever* ("I am warm" L). Even where the fever involves a rise of only one or two degrees, the result is great discomfort and a feeling of lassitude. So well does our thermostat work, that a fever is a sure sign of trouble, and a clinical thermometer is every mother's most prized piece of diagnostic machinery.

Perspiration has its unpleasant side too, for odor is associated with it. The ordinary sweat glands of the body generally are not at fault in this respect. There is a special variety of sweat gland, somewhat larger than the ordinary kind, which is concentrated in relatively few areas of the body, notably in the armpits and about the genital organs. These also secrete an odorless sweat but one containing small quantities of organic substances easily broken down by bacteria on the skin. The breakdown products are responsible for the characteristic human "body odor." Since these sweat glands become active only in puberty, children are relatively free of such body odor (although they can, of course, smell for a variety of other reasons).

Undoubtedly body odor must have been of use in primitive tribal days. It may have helped keep the tribe together in wooded areas or at night, where sight would not entirely serve. It may also have been a means of distinguishing a fellow tribesman from a stranger, since there would be slight variations in odor which the better developed sense of smell of early man might have picked up easily. It may even have served as a sexual stimulant.

In our own crowded society of today, however, in which each of us is placed in contact with hundreds or even thousands of strangers daily, and where odor is required neither for togetherness nor for counterespionage (though the matter of sex remains moot), odor has become a source of discomfort. Hence our modern emphasis on frequent bathing and on the use of soaps, perfumes, and chemical deodorants in an unending war against a natural phenomenon.

An important type of modified sweat gland is the milk-producing *mammary gland* ("breast" G). This would make milk a modification of sweat, which seems odd and even repellent, but this is no stranger than considering the larynx a modified gill bar or a leg a modified fin. The mammary glands are present only in mammals, and the very name of this class of chordates is derived from that fact. In general, they develop along two *mammary lines* down the ventral surface of the body. A number of the glands come together here and there along those lines to form a series of protuberances called *nipples* (AS). (The most primitive of living mammals, the duckbill platypus, lacks nipples, and milk that oozes out of the mammary glands must be lapped up by the young. In all other mammals the existence of a nipple makes it possible for sucking to take place, and this is a more efficient method for collecting the milk.)

In animals that produce a number of young at a birth, a number of nipples are retained along each mammary line. This can be easily observed among cats, dogs, and pigs. The cow, which usually brings forth one young at a time, retains but two pairs of nipples at the abdominal end of the mammary lines, and combines all four into one large baglike *udder* (AS). In the human being, which also brings forth but one young at a time ordinarily, only one pair of nipples is retained, toward the thoracic end of the mammary line, and these retain their separate structures.

In children and in adult males, the nipples remain small and without function; that is, they are *rudimentary* ("a beginning" L). About the time of puberty, however, the mammary glands of girls begin to change. The nipples enlarge, and the individual glands (of which there may be 15 to 20 on either side, each possessing an individual duct through the nipple) are surrounded and bound together by connective tissue and fat to produce a pair of soft, rounded *breasts* (AS). The individual female breast extends from the second to the sixth rib and from the breastbone to the armpit. The nipple is situated a little below the center of the breast, is pink in color to begin with, and is surrounded by a light pink area called the *areola* (a-ree'oh-luh; "small area" L). The nipple and areola are usually darker in brunettes than in blondes. After the first pregnancy, melanin is, for some reason, deposited in the nipple and areola, which consequently are more or less permanently darkened.

Milk is produced only after the birth of a child, and then only for as long as it is periodically withdrawn from the breast. Through evolutionary pressures, each species of mammal produces milk particularly adapted to the needs of its young. Human milk is 1.5 per cent protein, 7.2 per cent lactose, 3.6 per cent fat, and 0.2 per cent minerals. The rest is water.

Among animals generally, milk is a food of infancy only, to be abandoned forever after weaning (unless the adult animal is supplied with some by a human being). Even in prehistoric times man recognized it as a valuable food for adults, and kept cows, goats, sheep, even horses in order that these might be milked at such times as they were lactating.

Milk itself was difficult to keep fresh in the days before refrigeration became commonplace, and it was found best to direct its fermentation into pleasant channels to produce yogurt, or soured cream, or hundreds of varieties of cheeses. Milk fat was isolated as butter. Nowadays, of course, we use milk itself, kept uncontaminated by disease-producing bacteria through a preliminary process of gentle heating. This is called *pasteurization*, after Louis Pasteur, who introduced the process (in connection with wine, by the way, rather than milk).

Cow's milk, which is the variety drunk almost exclusively in the United States, does not have quite the same composition as human milk. It is only half as rich in lactose and is twice as rich in protein. (Calves grow at a faster rate than babies do and need more protein in their food.) Adults can tolerate this difference with ease, but infants cannot. For that reason, when human mothers are unable to (or choose not to) feed their infants at the breast, the cow's milk they use instead must be suitably modified. It must be diluted to keep the proteins from being too concentrated, and then because the carbohydrate content is brought even lower as a result, additional sugar must be added. The sugar added is usually in the form of dextrins (breakdown products of starch), which have the energy content of sugar but, like lactose, are relatively tasteless.

HAIR

The regulation of heat is aided in most mammals, as I have said earlier in the chapter, by a coat of hair. In addition to this primary use as a heat insulator, hair can be put to specialized uses. Crinkly hair with a pattern of overlapping scales that cause each to catch on its neighbor forms a mat or felt that is commonly called *wool*. Sheep are especially bred for this variety of hair.

Short, stiff, unusually thick hairs, like those on the back of swine, are *bristles*. The bristles in the walrus mustache may be up to a quarter of an inch in diameter. The spines of the hedgehog and the quills of the porcupine are hairs stiffened, pointed, and even barbed to serve as weapons of defense and offense. A still more extreme organ of defense born of the hair is the hornlike projection on the snout of the rhinoceros, which is formed by the fusion of a large number of hairs. Then, too, long, rather stiff hairs, richly supplied with nerve endings at the root, can be used as delicate organs of touch. Example are the long hairs forming the cat's mustache.

Such specializations are absent in man and, indeed, our coat of hair is a poor one altogether. In part this is a general primate characteristic, for the primates as a group are less hairy than are most nonprimates. This need not be surprising, for primates are tropical animals, among whom hair is not particularly important as an insulator. Then, too, the larger the animal the smaller the surface in relation to volume, by the square-cube law (see Chapter 5). For this reason it is easier for a large animal to stay warm because there is a large volume to produce heat and a comparatively small surface to radiate it away. It is for this reason that polar animals (such as the polar bear, the muskox, the walrus) tend to run to large size. Tropical animals when large in size must do something to improve the ability of the skin to radiate heat, and the easiest solution is to cut down on the hair cover or lose in entirely. The elephant, hippopotamus, and rhinoceros are all virtually hairless for example (though during the glacial ages woolly species of both elephant and rhinoceros roamed the north). The elephant, particularly the African species, has also developed large ears, which serve as heat radiators.

There is a similar tendency to lose hair among the larger primates. The gorilla, to take an example, has a hairless face and chest. In man the tendency has progressed the farthest, although he is smaller than the gorilla. This has not been in the direction of a true loss of hair. There are only restricted areas of the human body, such as the palms of the hands and the soles of the feet, that are truly hairless. As for the rest of the skin, hairs are numerous—as numerous per unit area as in other primates. In man, however, most of the hairs remain small and fine and do not grow thick enough or long enough to form a continuous cover or to trap an insulating layer of quiet air next to the skin.

Nevertheless we possess one ability that is reminiscent of ages past when our ancestors did possess a shaggy pelt of hair. Animals, under the stimulus of cold, can erect their hair, causing the coat to extend farther from the skin. In this way a thicker layer of air is trapped and insulation against heat loss is improved. We still possess small bundles of smooth muscle, *arrectores pilorum* (ar′ek-toh′reez py-loh′rum; "hair-erectors" L), which can tighten the skin and lift the hair. All we can do, however, is to erect a tiny useless hair while the skin at its base puckers and rises is "gooseflesh." Hair also stands erect in response to the stimulus of fright, as can be seen spectacularly when a cat comes unexpectedly face to face with a dog. The thicker coat of hair apparently serves to increase the dimensions of the animal and make it seem more formidable. We respond similarly to fright and develop gooseflesh.

Man has retained a true coat of hair, or at least patches of hair, in those areas where it will serve a protective function. The area where this is most marked is the top of the head. Here the purpose served is one of insulation, not so much against the loss of body heat as against the heat of the sun. There are brain proteins that are unusually unstable under even relatively mild degrees of heat, and the direct heat of the sun upon the exposed skull can, under severe conditions, produce unconsciousness and prostration, an effect called *sunstroke*. The coating of hair reduces the likelihood of this and the crisp even coating of frizzy, wool-like hair among many African Negroes is particularly efficient in this respect. Even with a coating of hair in full bloom, it is perhaps advisable for those not acclimated to the tropic sun to wear a head covering for further protection.

The various orifices leading into the body are usually protected by hairs. In addition to the hairs of the nostrils, mentioned in Chapter 5, there is usually a growth of hair about the ear canal and about the anus. The eye is fringed with eyelashes and, farther above, with an eyebrow. The former protect the eye against foreign particles, the latter against the glare of the sun.

The social custom among us which compels men and boys to cut the hair of their head short gives the illusion that female hair is longer than male hair. (In fact, it is my experience that American children of preschool age uniformly distinguish the sexes by the length of hair.) Actually this is not so, and in eras when it was permissible for males to let their hair grow they did as well at it as women did.

Hairiness increases with age. In some ways this increase is uniform among the sexes. At puberty, for instance, both males and females begin to grow hair in the armpits and about the genitals. The former is *axillary hair* (ak'si-ler'ee; "armpit" L); the latter *pubic hair* ("adult" L). Since the hair appears at the same time that the specialized odor-producing sweat glands do, it is natural to suppose that the function of the hair is to encourage odor-formation by serving as collectors for the bacteria producing it. This presents problems to our own odor-conscious society.

In other ways, the increase of hairiness with age is a particularly male phenomenon. In the later teens hair sprouts along the cheeks, chin, and upper lip to produce the beard and mustache. This cannot have any truly vital use (though it might serve to protect the mouth), since the female gets on very well without it. It might therefore be classified as a secondary sexual characteristic; one developed because it serves to advertise the sexual maturity of the male and to act as a stimulant to female susceptibility—like the mane of the male lion or the beautiful colors of most male birds. In addition, men develop more prominent and longer hairs on the shoulders and chest. This varies radically from group to group, so that men of European descent are considerably hairier, by and large, than other groups are. It also varies from individual to individual, so that some men are virtually bald-chested and others of similar descent are almost shaggy.

There is an opposing tendency that also varies radically from individual to individual: that of losing the hair on the head (though not elsewhere) in middle life and beyond. In some cases, signs of loss appear even in the twenties. A tendency to such *baldness* (AS), or *alopecia* (al'oh-pee'shee-uh; "baldness" G), is apparently inherited and is, in a small way, a sign of masculinity, since the tendency does not manifest itself, even though inherited, unless there is more than a certain critical concentration of male sex hormone in the blood. As a result, women rarely go bald, and men who are castrated before adolescence apparently never do.

With age there is also an increasing tendency for a loss of pigment in the hair. As the years pass, hair becomes more and more markedly gray, the process often beginning at the temples. Eventually it can turn completely white. As in the case of baldness, the process may start at quite an early age in particular individuals; and the tendency to premature graying is also an inherited one.

Hair is a product of the epidermis, which buckles deep into the dermis at the point where a hair grows. This deep dip of a thin layer of epidermis is the *hair follicle* ("little bag" L). Each follicle holds a single hair. Within a bulge at the bottom of the follicle is the *hair root*. This is alive. As hair is formed by the root, it pushes upward as a shaft covered with tiny, regularly placed scales and reaches the surface of the skin. The hair above the root is a nonliving structure consisting mainly of keratin.

Human hair grows at the rate of about 0.3 millimeters a day, which is equivalent to about an inch in ten weeks. Periodically a hair will fall out and a new hair will grow in its place. The frequency with which this happens will control the maximum length of hair. In certain animals, there are definite times of "shedding." This sometimes heralds a seasonal change, so that there is a growth of longer, thicker hair for the winter season. The new hair may even show a change in color: an animal may turn white for the winter—a condition that improves its chances of survival, since it is less easily seen against the snow.

In man, however, each hair has its own individual life cycle independent of its neighbor's. There are always some hairs falling out and being replaced and the thickness and texture of the hair (barring the advent of baldness) and the color (except for graying) remains constant.

Hair may be straight, wavy, or frizzy, but the exact reasons for this difference are not known. Since American fashion decrees that women ought to have wavy hair, the prevalence of straight hair among them is a source of dismay to the woman and of income to beauty parlors. The keratin molecules of hair are bound together by atom-chains that can be broken by damp heat and the action of certain chemicals. If the hair is then curled and left so for a period of time, new atom chains form in the new position. The hair will be held in a permanently curled position then —a "permanent wave," in fact—as long as it remains in the head. The new hair formed at the root will be straight, and after it falls out an artificially curled hair will be replaced by a straight one. A permanent wave is therefore only temporarily permanent.

Associated with virtually every hair is a small *sebaceous gland* (see-bay'shus; "tallow" L), which produces *sebum* (see'bum), a waxy secretion. A duct leads this to the hair follicle. The hair and surrounding skin is constantly coated with sebum, unless it is painstakingly soap-and-watered away. Sebum contributes to the glossiness of hair and acts as a protective and waterproofing agent to both hair and skin. Hair so protected sheds water. It has been pointed out that the natural direction in which hair lies on the human body is such that if a man squatted with knees pulled up to his chin and with his hands behind his neck, elbows pointing downward (a position a primitive man might well take up in a miserable attempt to reduce exposure to a pelting rain), all the body hairs would point downward, shedding rain.

CROSS SECTION
OF SKIN

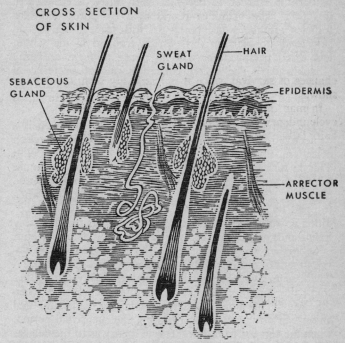

A woman's nose is particularly well supplied with sebaceous glands. It is sebum that produces the "shiny nose" against which women, armed with powder puffs, wage constant war. It is collected sebum that forms earwax and the matter that accumulates in the corners of the eyes after sleep. It is also sebum, rather than perspiration, that makes a person's hair and skin markedly greasy after a period without washing.

On the other hand, washing, if carried to an extreme, can unnecessarily deprive hair and skin of a useful protective agent. It should be pointed out that lanolin, which is widely touted by advertisers as a component of various

preparations designed to help the hair and skin by re-
placing the sebum lost through overapplication of various
other preparations as widely touted by the same advertisers,
is itself a product of sheep sebum.

Despite all natural protective devices the skin, particularly if allowed to become overly dirty, is exposed to
infection. The openings represented by pores and hair
follicles are the weak points in this respect. Unwashed
skin tends to pick up dirt, which, trapped in the sebum
and dried perspiration, will fill pores to form unsightly
"blackheads." The richer concentrations of bacteria on
dirty skin are likely to start an infection at the site of the
blackhead or in a hair follicle to produce the inflamed
areas we call pimples and boils.

The activity of some microorganisms is apt to elicit a
protective response in the form of an exaggerated secretion
of sebum, or *seborrhea* (seb′oh-ree′uh; "sebum-flow" L).
This in turn causes or is accompanied by skin inflammation (*dermatitis*) and itching. The sebum collects and
produces a greasy scale that is most noticeable when
trapped by and accumulated in hair. We know this condition as *dandruff* (AS), which is unsightly, uncomfortably
itchy, and, in extreme cases, can bring about loss of hair.

Seborrhea seems to increase the likelihood of the development of chronic pimple-formation, or *acne* (ak′nee).
There seem to be hormonal imbalances involved here, because the condition appears most often at puberty, when
the body's hormonal pattern is shifting radically, and
among girls is most marked during the menstrual period,
when the same is true. (In fact one theory as to the derivation of the word "acne" is that it was originally a misprint
of "acme," and that it was meant to indicate that the
condition struck at the height of youth, the acme of life.)
Probably no other disease of so mild a character has such
serious psychological effects, for the young man or woman
is usually disfigured at just that time of life when he or
she has discovered the opposite sex and is most insecure
and self-conscious about the whole matter. Furthermore,
although acne is usually a passing phase and tends to
vanish with the teens, it leaves scars behind on the face as
well as the personality.

The skin is also subject to fungus infections, of which

the best known is "athlete's foot," and subject to itchy, scaly disorders that don't always have a recognized cause or a secure treatment—such as *eczema* (ek'zi-muh; "eruption" G), *psoriasis* (so-ry'uh-sis; "itch" G), and *impetigo* (im'peh-ty'goh; "to attack" L). Rashes and pustule formation are common in many diseases like measles and chicken pox. Next to the clinical thermometer, the mother's best diagnostic aid is the eye with which she scans her child's body for any telltale rash.

Chicken pox may result in scarring, but the real villain in this respect is the now happily conquered smallpox. The real terror of smallpox in the days before vaccination was not so much the possibility of death—which, after all, accompanied almost any other disease in those medically ignorant days—as the possibility of continuing life with so scarred a face that every vestige of good looks, and almost of humanity, was gone.

The skin may contain local heterogeneities. If these are pigmented and darker than the surrounding skin, they are *moles* (AS). Sometimes the disfigurement is caused by a small knot of dilated blood vessels in the skin and from its color and appearance it is then called a "strawberry mark." When moles are present at birth, particularly if they have an irregular shape, they are called "birthmarks." Most people have small moles and are generally sufficiently accustomed to them to be unaware of them. Moles, however, can occasionally become cancerous, and although this does not happen often, the possibility makes it advisable for individuals to note any changes in the appearance of a mole and to consult a doctor promptly in this connection. Warts are not merely large moles, but are apparently the result of a virus infection. Although disfiguring they are not dangerous.

Still, though the catalog of skin disorders is long and although man is more aware of them than of disorders in other organs, simply because it is the skin that is exposed to view, it should be emphasized that on the whole skin is an extremely effective shield. It performs its guardian duties well and the wonder is not that it occasionally gives way somewhat to the buffets of the environment but that it gives way so infrequently and has such a remarkable talent for recovery.

11

OUR GENITALS

REPRODUCTION

Despite the beautiful and awesome adaptation of the human body to its functions, it is at best but a temporary structure. Even if the body were—by luck, by care, by innate health—to avoid the infectious diseases and maladjustments I have mentioned here and there in this book, it cannot go on forever. There are inevitable changes that come with age, a hardening and loss of resilience of the connective tissue and of the artery walls; a slow deterioration and loss among the nerve cells; tiny losses of essentials and tiny accumulations of wastes here and there. We are not certain what the exact nature of the fundamental changes with time may be, but, taken together, they make up the aging process.

The result is "old age." This is not a strictly chronological matter, for some human beings retain alertness and a measure of vigor into the nineties and others begin to dodder in their sixties. Medical science has not yet learned how to prevent, or even slow, the fundamental aging process; and although the conquest of many infectious diseases and the successful treatment of some metabolic disorders have raised the life expectancy over recent centuries from 30 to 70, the maximum age which may be reached remains at a little over a hundred.

And in the end, inevitably and universally, comes death.

The human body as I have described it in previous chapters must yet contain some organs I have not described, for if human life is to continue while individual bodies die, provision must be made for the formation of new bodies at a rate at least equal to that at which old bodies die. This process of *reproduction* is fundamental to all living things.

A one-celled creature can reproduce itself simply by dividing in two;* that is, by *binary fission*. Among the simpler multicellular organisms, binary fission is sometimes retained as a method of reproduction. Some coelenterates can divide along a longitudinal plane (from head to foot) and form two of themselves. Some simple worms can divide along a transverse plane (from side to side) and form two of themselves; or may fragment (*multiple fission*), each piece that results giving rise to a separate organism.

Such fission, binary or multiple, requires that each fragment be able to regenerate the rest of the organism. This can be done when the organism is a single cell or when it is made up of relatively few and unspecialized cells. However, it seems a rule that the more specialized a tissue the less regenerative it becomes. In the human being, tissues as specialized as nerve and muscle can regenerate little, if at all. It is reasonable, then, to expect that as the cells of a given organism grow more numerous and specialized, reproduction by binary fission becomes more and more impractical. Fission must therefore give way to more sophisticated methods of reproduction, or a quick

---

* The use of the word "simply" is misleading—there is nothing simple about the process. The chemical functioning of the cell is guided by structures in the nucleus (called chromosomes) which are made up of a series of molecules of deoxyribonucleic acid, a name usually abbreviated as DNA. When a one-celled creature divides, the various DNA molecules form replicas of themselves in order that, after cell division, each "daughter cell" possess an identical set of its own. The process by which this is accomplished is only becoming understood in the years since 1950, and this is described in considerable detail in my book *The Wellsprings of Life*.

limit is placed on the permissible complexity of multicellular life.*

One way out is to restrict the reproductive process to a small region of the organism which remains relatively unspecialized, and is therefore capable of regenerating all the rest of the organism. The beginnings of such a process can be traced back to one-celled forms of life, for yeast cells do not multiply by dividing evenly in two, but form small outgrowths of protoplasm at restricted regions of the cell boundary. These outgrowths grow larger and finally break away as full cells. This is *budding*. In the yeast cell there seems scarcely any way to distinguish this from binary fission, but in multicellular animals, the analogous process represents a clear specialization. The fresh-water hydra, a coelenterate, will grow a small group of cells at one point on its surface and these will multiply and form a new hydra that will eventually break off. Only some cells, rather than all, are involved.

The end toward which this specialization aims seems inevitable. The logical result is for an organism to produce a single cell, supremely specialized through its very lack of ordinary specialization, designed for the sole purpose of regenerating an entire organism like the one that produced it in the first place. Such a cell is an *egg cell*, or *ovum* ("egg" L).

So far, reproduction as I have described it seems to be the function of a single organism, but it is only in the simplest forms of life that this is true. Among the organisms most familiar to us, each species is divided into two groups or *sexes* ("to divide" L) and it is as a result of the combined activity of one of each sex that new individual organisms are produced. Reproduction on the basis of the activity of two separate individuals is therefore *sexual reproduction*, whereas fission and budding are examples of *asexual reproduction*.

Sexual reproduction can be traced back to the more complex one-celled creatures. Among these a period of reproduction through binary fission seems slowly to reduce

* Nevertheless, the ability to reproduce by binary fission is never entirely lost, no matter how complicated an organism grows. The individual cells of all organisms, including ourselves, if they are capable of multiplying at all, multiply by binary fission. Therefore, although man does not reproduce by fission, he grows by fission.

the vigor of the individual. When this happens, two of the tiny creatures not closely related (that is, they have not formed from a single ancestral cell of a relatively small number of generations before) meet. The cell membrane between them breaks down over a limited area, and the two interchange portions of their nuclei. This is *conjugation* ("yoke together" L), and the process restores cellular vigor.

Why this should be is not certainly known, but one can speculate that, after a period of asexual reproduction, tiny errors accumulate in the replication of the chromosome material and that this may slowly lessen the efficiency of the chemical machinery of the cell. Through conjugation, each individual receives part of the DNA of the other, and this dilutes out the changes for the worse. One individual is likely not to suffer the particular evils of the other, and the strengths of each negate the weaknesses of the other. Furthermore, any accidental change in the chromosome material that may be for the better is spread by means of conjugation from one group of individuals to the remainder.

In general, this would seem to exemplify the purpose of sexual reproduction. Through this process there is a mixing of the mechanism of inheritance of two individuals, allowing new combinations to develop in each generation and spreading changes developed by one to all the others of a species. This apparently hastens the evolutionary machinery and in the end seems to work for the good of the species.

The ultimate proof of this lies in the changes we observe in the course of evolution. Only the simplest organisms reproduce by asexual mechanisms only. The more specialized one-celled organisms reproduce by asexual mechanisms, with an occasional sexual step (conjugation) introduced. In many of the simpler multicellular animals there is an alternation of generations. That is, there is first an asexual reproduction, producing individuals capable of sexual reproduction; these in turn produce, by sexual mechanisms, individuals that reproduce only asexually. In the more complex multicellular animals, including ourselves, however, asexual reproduction is dropped entirely and sexual reproduction becomes the norm.

Once an organism reaches the stage where it can produce an egg cell, it is almost invariably involved in sexual reproduction. It is possible, to be sure, for an egg cell by itself and without interference of any other cell, to form an organism. This is called *parthenogenesis* (pahr'theh-noh-jen'eh-sis; "virgin birth" G), and it can take place among numerous invertebrates. Among the bees it takes place routinely, for example. Taking life as a whole, though, this is unusual, and among vertebrates it never occurs.

In general, an egg cell will not produce an organism until it is fused with another type of cell called a *sperm cell* ("seed" G), or *spermatozoon* (spur'muh-toh-zoh'on;

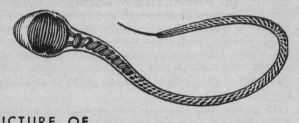

STRUCTURE OF SPERMATAZOON

"animal seed" G). This process of fusion of egg and sperm is *fertilization,* and the combined cell, now capable of forming a complete organism, is a *fertilized ovum.* In a number of simple animal forms, the same individual can produce both egg cells and sperm cells. Theoretically, it is then possible for a sperm cell to fertilize the egg cell of the same individual. This defeats the entire purpose of sexual reproduction, however, and such an animal is usually so designed that this form of self-fertilization cannot take place. Thus, an earthworm produces both egg cells and sperm cells. But reproduction comes about only when two earthworms make contact in such a fashion that the sperm-producing region of each is next to the egg-producing region of the other. Each earthworm fertilizes the other and not itself.

The natural step that would outlaw all possibility of self-fertilization and make certain that the purpose of sex is not thwarted is to arrange for individuals to produce only egg cells or only sperm cells, never both. The egg-producing individual is the female and the sperm-producing one the male. All vertebrates, including man of course, are divided in this way into a male sex and a female sex.

THE EGG

When a one-celled animal divides, each daughter cell is already large enough and complicated enough to continue independent life on its own. It is virtually born an adult. The fertilized ovum of an even fairly simple multicellular creature must, however, undergo a number of cell divisions before it becomes large enough and specialized enough to be able to assume an independent existence. During this period of development, the cell mass requires energy and, since it cannot eat, it must have a pre-existing food supply. The egg cell therefore contains food material or *yolk* (yoke; "yellow" AS, which describes its usual color) in sufficient quantity to bridge the gap between fertilized ovum and an organism complicated enough to feed independently. The presence of the yolk makes the egg cell larger than ordinary cells and all the more immobile by virtue of that fact.

If a fertilizing union between sperm cell and egg cell is to take place, then, the full responsibility of the meeting of the two must lie with the sperm cell. The sperm cell must therefore be capable of motion. To attain this end, the sperm cell is made light by being supplied with virtually no food. It need only live long enough to travel from its point of release to the point where the egg cell lies. If it doesn't succeed in that one task its reason for existence is gone and there is no point in feeding it further. Nor need it carry any cytoplasm; the egg cell will have enough for both.

In short, the sperm cell need carry nothing but the nuclear material within which lies the inheritance machinery. For that reason the sperm cell is much smaller than the average cell, and a veritable pygmy in comparison with the yolk-bloated egg cell. Nevertheless, the two cells, the tiny sperm and the giant egg, contain equal amounts of the nuclear material and both contribute equally to the inheritance of the new organism.

To move the tiny sperm cell a flagellum is all that is needed. By use of it, the sperm cell, looking for all the world like a microscopic tadpole, lashes it way through

**FERTILIZATION
OF OVUM**

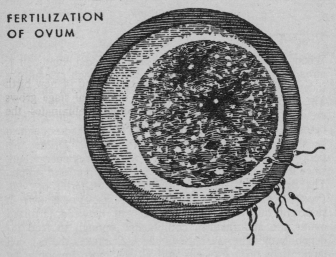

the water to the egg cell—provided, of course, it finds itself in a watery medium.

Fertilization for sea animals, immersed as they are in a watery medium, can, therefore, be simple enough. A female of the species will deposit a mass of eggs in some secluded spot, and when she has finished a waiting male will swim over the eggs, emitting a stream of sperm cells. The sperm cells move in the direction of the egg cells and the two parents swim away, their work done.

Among land creatures matters are more complicated, since sperm cells cannot swim through air or along the ground. It is therefore necessary that the male deposit the sperm cells, plus fluid, within the duct through which the eggs will eventually emerge. The sperm cells will reach the egg cells within the body of the female and when the eggs are later laid, they will be already fertilized. This process of fertilization within the body is called *copulation* ("to couple" L), and for the purpose the male has developed a copulatory organ which can be inserted into the female.*

The helplessness of an egg during the time in which it is developing into an organism makes it desirable from the standpoint of survival that it reach the independent organism stage as quickly as possible. Yet the more complex the phylum, the longer it takes for development to full complexity to take place. In many cases a compromise is struck; the egg is rushed to the stage of independent organism before full complexity is reached. Such a simplified but independent organism is a larva. The larva feeds and grows until it is ready to undergo a kind of new birth and become an adult. (Sometimes the larval stage grows more and more important and actually eliminates the adult form, as I explained in Chapter 1.)

We are most familiar with this phenomenon in the case of certain insects, for we all know the caterpillar which at a certain stage of its life cycle makes what is almost a second egg for itself, in the form of a cocoon, and emerges as the adult butterfly. Among the vertebrates the frog is the best-known example, for it hatches as the larval tadpole and later gradually becomes a frog, without, however, immobilizing itself in a cocoon stage in the process.

Even so, there is a tremendous mortality among larvae, which are too small and simple to be as successfully self-defensive as the adult of the species. To ensure survival of the species, the most common device is therefore to lay eggs by the millions, in the hope that out of all those millions at least one or two will survive to adulthood by sheer good fortune. This actually happens; otherwise most species of fish would be extinct by now.

The alternative is to adopt some form of protection for the eggs and larvae till they can fend more successfully for themselves. Thus, some fish build nests or keep the eggs in pouches or in their mouths until development has well progressed. There are even cases where fertilized eggs remain within the body cavities of the female until the

* Human beings, like other land animals, must engage in copulation if reproduction is to take place. However, because there is a certain shamefacedness about discussing the process which, in our culture, is kept very private, it is referred to by the rather prissy term *sexual intercourse.*

larval stage is reached. These alternates are poorly developed among sea creatures, however.

Among land vertebrates the care of eggs becomes more important for several reasons. In the first place, the larval stage must be eliminated on land. The land, as an environment for life, is unbelievably hard and cruel as compared with the kindly sea, within which life was first developed. The force of gravity is uncanceled by buoyancy on land and it must be combated by fully developed muscles; the atmosphere must be breathed by fully developed lungs; and water conserved by a fully developed kidney. The young reptile therefore must emerge from the egg, smaller than the adult perhaps, and weaker, but not essentially simpler in structure. It is always recognizably of the adult form and is capable of independent life in the fashion of its parents.

To eliminate the larval stage means a lengthening of development within the egg, which in turn means that the egg must contain a large quantity of yolk. Consequently, each egg represents a tremendous investment, and because of its size there is a sharp limit to the number that can be laid. Where fish and frogs can lay vast numbers of eggs and trust to luck that some will survive, reptiles and birds must lay small numbers and take some care of them in consequence. Reptiles may bury their eggs in sand or even remain in the vicinity until hatching time, but birds are the classic examples of creatures that care for their eggs.

The care of eggs is made all the more necessary in that birds are warm-blooded. This is something of great advantage to survival, for it means that the birds' body mechanisms proceed at the same rapid rate under all conditions of outward temperature. Birds are as busily energetic in the chilly morning as in the hot afternoon and need not, like the reptiles, wait until the sun warms their sluggish cold-immobilized muscles. Birds can even survive the cold of the winter without trouble as long as they can find food.

But warm-bloodedness has a disadvantage in that it must be maintained. Bird tissues can no longer endure the chilling that reptilian tissues can. This temperature-maintenance can be handled by the bird while it is an adult, as long as it is alive at all, but what about the developing bird within the egg? It is too small to produce heat at a rate great enough to match the radiation of heat from the egg surface. Left to themselves, bird eggs would not hatch. This is why the parent birds (usually the mother, but sometimes the father as well) must incubate the eggs with the warmth of the adult body. It is a tedious job, but it ensures the survival of all or almost all the eggs. Even after the birth, the parent birds often undergo a long and patient ordeal during which they must feed the perpetually yawning beaks of the still helpless young.

An alternative to the complex care of egg and young as displayed by the birds arises through tendencies already present among the reptiles and even among the sharks. Sometimes the fertilized egg is retained within the ducts that would ordinarily lead them to the outer world and kept there until they hatch. It is then the living young that are "laid" rather than the egg. Whereas an egg-laying creature is said to be *oviparous* (oh-vip'uh-rus; "to bring forth eggs" L), those which retain eggs and allow the living young to issue are *ovoviviparous* (oh'voh-vy-vip'uh-rus; "to bring forth eggs alive" L). The eggs of an ovoviviparous animal are sure of survival as long as the mother survives, and this is itself important. Even more important for us is the fact that it gained a subsidiary use. When a particular group of reptiles developed warm blood and became the first primitive mammal, the retention of eggs within the ducts served not only as protection but also as a means of keeping them warm. Ovoviviparity became a kind of internal incubation.

THE PLACENTA

The earliest mammals did not completely develop the ovoviviparous mechanism. We know that because some of them have survived to this day, although only in Australia and New Guinea. These lands had split away from Asia before higher and more efficient forms of mammals had developed. Secure from competition with the late-comers, early mammals survived in this out-of-the-way continent. Elsewhere on the earth they did not. When early explorers of Australia returned with reports of animals

with hair (and therefore mammals) that nevertheless laid
eggs, they were scoffed at, but they turned out to be telling
the truth. These egg-laying mammals are placed in a separate subclass by themselves, the *Prototheria* (proh'toh-
thee'ree-uh; "first beasts" G). Their eggs have not quite
finished developing at the time they issue from the body
and so they must be incubated during the final stages of
development.

The Prototherians show other primitive characteristics
too. They are imperfectly warm-blooded; and although
they produce milk, one of the species of these beasts, the
platypus, lacks nipples. Furthermore, the urethra, rectum,
and birth canal of the Prototherians all empty into a final
common channel with one opening to the outside world.
This single channel is the *cloaca* (kloh-ay'kuh; "sewer"
L). The cloaca exists among the other vertebrate classes
generally, but it is unusual for a mammal to have one. All
mammals except the Prototherians have more than one
opening from the pelvis to the outer world. The various
living members of Prototheria are therefore classified in the
single order of *Monotremata* (mon'oh-tree'muh-tuh; "one
hole" G).

A second subclass of mammals are the *Metatheria*
(met'uh-thee'ree-uh; "middle beasts" G), and these are
ovoviviparous. The egg is reduced in size and the time
of the development within the egg is consequently shortened. The young hatch out before the egg is laid, so living
creatures issue from the birth canal. The difficulty here,
though, is that the shortening of the time of development
leads to the birth of young too undeveloped to be capable
of leading an independent existence. They have just the
strength to work their way through the maternal hair to
an abdominal pouch and to climb within. There they
fasten on to a nipple and remain affixed while they complete their development. What the egg left unfinished the
pouch finishes, and it is almost a return to the larval system. Because of the presence of the pouch, all species of
this subclass are lumped into the single order *Marsupialia*
(mahr-syoo'pee-ay'lee-uh; "pouch" L).

Though the marsupials were more successful than the
monotremes and survive today in a larger number of
species, the device of the pouch is inferior to what was

to come, so that almost all the marsupials (the kangaroo is the most familiar example) are, like the monotremes, confined to Australia and to nearby islands. The only marsupials outside that area are the opossums of the Americas, and they survive only through sheer fecundity.

In fact, although mammals first developed comparatively early in the Age of the Reptiles, they remained quite unremarkable little creatures for tens of millions of years, during which time reptiles dominated everywhere. The scurrying mammals gave no signs of being the future lords of the planet, and one of the reasons for it perhaps was that they were all members of either Prototheria or Metatheria, or of a third undistinguished and primitive group which is now extinct.

It wasn't until after the climatic changes heralded the end of the Mesozoic that a final group of mammals, the only one which in the long run proved truly successful, was developed. These were of the subclass *Eutheria* (yoo-thee'ree-uh; "true beasts" G). Among the Eutherians, the Metatherian advance was carried on to its ultimate end. The egg was made still smaller until, indeed, it was down to pinhead size, smaller even than most fish eggs. It was impossible for any creature as complex as a mammal to develop out of such an egg with enough life in it even to get to a pouch. The solution was therefore to leave it within the body and feed it there. In other words, the developing Eutherian egg attached itself to the wall of the birth canal, which developed an extensible pouch, or *womb* (AS), designed to hold the developing egg or eggs and to grow the developing young.

The membranes that surround all amniote eggs were modified among the Eutherians into an organ that hugs the inner surface of the womb. This is the *placenta* (pluh-sen'tuh; "cake" G, because it is flat and broad, like a pancake). The marsupials, some reptiles and sharks, and even a few nonvertebrates, develop organs with placenta-like functions in the course of the development of the eggs, but only among the Eutherians does it reach full development; and we ourselves, of course, are included among the Eutherians. The placenta is well infiltrated with blood vessels developed by the embryo,* while the wall of the birth canal is equally well supplied with blood vessels developed by the mother. There is no direct connection, to be sure. No actual blood flows from mother to embryo or vice versa. However, glucose, oxygen, amino acids— all the necessities of life—diffuse from the mother's blood vessels into those of the embryo. On the other hand, carbon dioxide, urea, and all the wastes formed by the embryo diffuse across into the mother's bloodstream.

Two arteries lead from the embryo through a narrow *umbilical cord* (um-bil'i-kul; "navel" L) to the placenta, where they break up into numerous capillaries. These collect again into a vein that leads the blood back through the umbilical cord. Naturally, it is the vein that, in this case, carries arterial blood, made so with oxygen taken from

UMBILICAL CORD AND SECTION OF PLACENTA

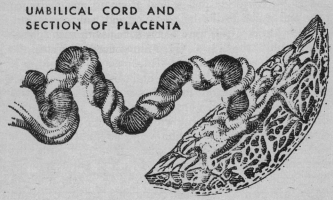

the maternal blood supply. The arteries of the umbilical cord carry venous blood. (This is analogous to the situation involving the pulmonary artery and veins in the free-living human being [see Chapter 6], and logically so, since in the embryo the placenta serves the same function that the lungs serve in the free-living human.)

The developing young has a form of hemoglobin somewhat different from the ordinary. This different form is

* *Embryo* ("to swell within" G) is the common term for developing young before it has reached the stage of independent existence. Among vertebrates, it is used to designate the earlier stages only, the later stages being designated by the term *fetus* ("fruitful" L). In the case of the human, the developing egg is an embryo for the first three months and a fetus thereafter.

hemoglobin F (F for "fetus") while ordinary hemoglobin is *hemoglobin A* (A for "adult"). Hemoglobin F forms a tighter union with oxygen than hemoglobin A does, so that oxygen is, so to speak, torn from the mother's hemoglobin A by the fetus's hemoglobin F and diffuses readily across the membranes from mother to fetus, but not vice versa. Hemoglobin F begins to be replaced by hemoglobin A even before birth, and the process is completed by four months after birth.

When the young issue from the womb, the placenta and other membranes emerge also, as the "afterbirth." The umbilical cord is severed but the place to which it was once attached is clearly visible in the human being as the *umbilicus* or *navel*.

The placenta-equipped Eutherians are sometimes called the *placental mammals*. Sometimes, also, the Eutherians and the Metatherians are combined into a single subclass, *Theria* (thee'ree-uh; "beasts" G), and are considered separate groups within that subclass, but with such refinements of classifications we need not concern ourselves.

THE HUMAN FEMALE

The organs involved in the process of reproduction may be lumped together as the *genitals,* or *genitalia* (jen'i-tay'lee-uh; "to give birth" G). Of these the most important are those which form the egg cells in the female or the sperm cells in the male. These are the *gonads* (gon'adz; "generator" G); in addition to the sex cells they produce hormones which govern the changes that take place in the human body at adolescence and maintain the reproductive system at working efficiency.

The female gonads, in which the ova are formed, are, very naturally, termed the *ovaries*. In the adult human female the ovaries are a pair of small organs, about the shape of a large but flattened olive, being 1½ inches long by 1 inch wide and ½ inch thick. They are formed in the kidney region, to begin with, but during the course of fetal development they descend to the pelvis, where they remain somewhat more ventral than the rectum but more dorsal than the bladder.

The ovary contains up to 400,000 potential egg cells, each located in a *primitive Graafian follicle* (named for a 17th-century Dutch anatomist, Regnier de Graaf). At the beginning of adolescence, when important hormonal changes begin sweeping over the female, a follicle-stimulating hormone is produced by the pituitary gland. This hormone brings about the maturing of the ova within the ovary, but usually one at a time and ordinarily at four-week intervals. This continues for something over thirty years, and in that interval perhaps 400 mature ova are formed, one for every thousand potential ova. (A number of follicles will develop simultaneously, but usually only one will carry through to completion, the others degenerating after their false start.)

The potential ovum, also called an *oocyte* (oh'oh-site; "egg cell" G), together with its follicle, grows and develops through a series of complex stages until it is larger than any other cell in the human body. After all, even though the embryo is to be nourished by its mother through the placenta, it must be supplied with enough of a yolk to carry it through the very early stages of development, during which it attaches itself to the wall of the womb and develops a placenta. Yet the ovum, though the largest human cell, is not really very large in an absolute sense. It is about 1/120 of an inch in diameter, which makes it just visible to the naked eye as the barest speck. About ten days after the beginning of the development of the oocyte, the follicle, which has become large enough to be sticking out of the ovary like a tiny blister, breaks open and the ovary is discharged into the body cavity. This is *ovulation*. The ruptured follicle fills first with blood and then with a yellow fatlike material. It is then called the *corpus luteum* (kawr'pus lyoo'tee-um; "yellow body" L).

Near each ovary is a duct, about 5½ inches long, called the *Fallopian tube* (named for the 16th-century Italian anatomist Gabriello Fallopio). It flares out near the ovary into a wide fringed opening (called *fimbria* from a Latin word for "fringe"). The ovum as it bursts out of its follicle is caught in this opening. Slowly the ovum makes its way down the tube. The tube is lined with cilia that through their motion sweep the ovum along. During its stay in the Fallopian tube, the ovum goes through the final stages of

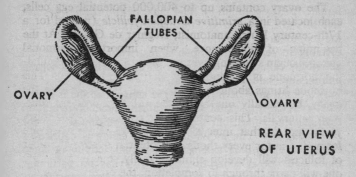

FALLOPIAN TUBES

OVARY

OVARY

REAR VIEW OF UTERUS

its development. It is also in the Fallopian tube where it meets the sperm cells if any are around. They will be there, of course, only if copulation has taken place at the time of ovulation. If copulation has not taken place, the ovum remains unfertilized and, after a day or two of waiting, dies.

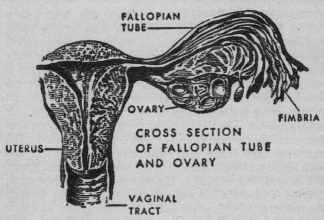

FALLOPIAN TUBE

OVARY

FIMBRIA

CROSS SECTION OF FALLOPIAN TUBE AND OVARY

UTERUS

VAGINAL TRACT

Since a woman usually ovulates only once in four weeks, this means she is only fertile a couple of days a month.

This waiting of fertilization on the chance of copulation is more marked among human beings than among most mammals. In many mammals, ovulation takes place only at certain times of the year, and during those times the female invites copulation. The male of the species is himself brought to eagerness through hormonal changes and the copulation that then takes place leads almost inevitably to fertilization. Such a restricted time of sexual activity is called a "rutting season" and during this season animals are said to be "in heat."

A woman may be in heat in the sense that she is somewhat readier for sexual activity at the time of ovulation than at other times, but there is no true rutting season for *Homo sapiens*. The individual human being is capricious enough to be willing to engage in sexual activity at some times and not at other times, but this has no particular relation to the calendar. Under the proper circumstances, both man and woman can engage in sexual intercourse at any time. This reduces the efficiency of copulation as a means of fertilization, and sexual activity often goes for naught as far as the production of young are concerned. However, this inefficiency is clearly not serious, in view of the human birthrate and the manner in which the human population is increasing.

The Fallopian tubes, one on either side, lead into a hollow pear-shaped organ in the midplane of the body, just above the bladder. This is the womb, or *uterus* (yoo'-tuh-rus; "womb" L). The uterus has strong muscular walls and a mucous membrane as an inner lining, one that is well equipped with blood vessels. This inner lining is the *endometrium* (en'doh-mee'tree-um; "inner womb" G). As the ovum makes its way down the Fallopian tube, the corpus luteum it has left behind produces a hormone that prepares the endometrium for the oncoming ovum. The endometrium becomes soft, moist, and even better supplied with blood vessels. When the ovum enters the uterus it will, if it is fertilized, adhere to the endometrium and begin to develop a placenta. It will remain in the uterus through the full term of embryonic and fetal development and during that time the uterus will grow along with the developing child. The increase in size becomes very noticeable in the latter half of pregnancy, as we all know.

On rare occasions more than one ovum ripens at one time, and if sperm cells are waiting each is fertilized by a separate sperm cell. In this way, if two ova are fertilized, *twins* develop and are born; *triplets* if there are three, and

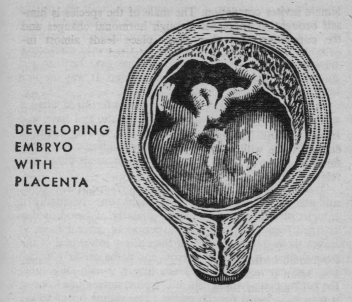

DEVELOPING
EMBRYO
WITH
PLACENTA

so on. Two children born simultaneously of separate ova fertilized by separate sperm cells are *fraternal twins*. They have each their own pattern of inheritance and resemble each other no more closely than do ordinary brothers and sisters born at separate times. Fraternal twins need not even be of the same sex. It may also happen that a single fertilized ovum in the course of its development will break up into two halves, each of which may proceed to develop and form complete organisms. (This is a form of asexual reproduction that can take place even in man.) Since these two individuals stem from a single ovum fertilized by a single sperm cell, they share the same pattern of inheritance. They are *identical twins*, always of the same sex and always very similar in form and feature. There may also be identical triplets, quadruplets, or even quintuplets.

It has been estimated that fraternal twins outnumber identical twins three to one. It has also been estimated that one out of 85 births are twins, one out of 7500 are triplets, one out of 650,000 are quadruplets, and one out of 57,000,000 are quintuplets. The greater the number of children at a birth the smaller each must be, on the average, and the less advanced in development. The chances of survival are less as the number at a birth increases, therefore. This accounts for the world excitement back in 1934 when Mrs. Oliva Dionne gave birth to girl quintuplets (identical), all of whom lived. It was the first such case in recorded history.

If, on the other hand, the ovum is not fertilized when it enters the uterus, it will quickly degenerate, as I have said earlier. The corpus luteum fades out also, leaving a scar on the ovary. (With age and repeated ovulations, the ovary loses its original smooth exterior and becomes pitted and uneven through scar formation.)

With the corpus luteum gone, its hormone is also gone, and the endometrium begins to break up. Eventually it sloughs off and, together with a quantity of blood, is discharged from the body over a period of several days. It leaves through the opening at the narrow lower end of the pear-shaped uterus. This narrow end is the *cervix* ("neck" L). The cervical opening leads into a 3-inch-long tube, the *vagina* (vuh-jy'nuh; "sheath" L, for, during intercourse, it serves as sheath for the penis). The vagina opens to the outside world just behind the much narrower urethra and somewhat before the anus. The sensitive area about these three openings is the *perineum* (per'i-nee'um; "around the wastes," in reference to the excretory openings).

The vaginal and urethral openings are enclosed, when the thighs are in the ordinary position, by two folds of tissue, the inner being the *labia minora* (lay'bee-uh minaw'ruh; "smaller lips" L) and the outer the *labia majora* (muh-jaw'ruh; "larger lips" L). The vaginal opening and the labia enclosing it are sometimes spoken of as the *vulva* (vul'vuh), from another Latin term for "womb." At the anterior end of the vulva, between the two labia minora, is a small organ about an inch long, richly supplied with nerve endings and very sensitive. It is the *clitoris* (kly'tuh-ris or klit'uh-ris; "to enclose" G, perhaps because it is enclosed by the labia).

It is through the vagina that the contents of the uterus are discharged. Where successful fertilization has taken place, it is a living child that ultimately comes forth. Where

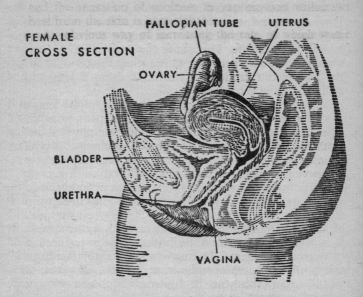

FEMALE CROSS SECTION

FALLOPIAN TUBE — UTERUS — OVARY — BLADDER — URETHRA — VAGINA

fertilization has not taken place, it is the degenerating endometrium that must do so in a slow bloody discharge that lasts for some days and is referred to as the *menses* (men'-seez; "month" L) or the *menstrual flow*. When that is completed the cycle begins again, and a new ovum begins to mature.

The cycle and its accompanying menstrual flow begins at about the thirteenth year, the first appearance of the menses being the *menarche* (meh-nahr'kee; "beginning of the monthly" G). It continues at monthly intervals, except where interrupted by pregnancy, until the age of 45 or 50, at which time it becomes increasingly irregular and ceases. The period during which it ceases is the *menopause* ("ending of the monthly" G). The whole process can be an uncomfortable one for the female. The menses are sometimes painful and may be preceded by a period of nervous tension and depression. It is occasionally difficult, during the time of menstruation, to engage in the ordinary business of life without precautions that are sometimes uncomfortable and usually troublesome, particularly since social custom demands that the whole matter remain unmentioned and ignored. The menarche itself often comes as a frightening experience to a young girl who may not be sufficiently prepared for it; and the menopause may involve months and even years of unpleasant symptoms while the body adjusts itself to the interruption of a long continued hormonal cycle.

The supposed greater emotional instability of the female as compared with the male may arise, in part, from the difficulties brought on by this cycle and the physical and temperamental upsets involved. It is a condition that is all the more frustrating to women in that its existence is either unsympathetically dismissed by men or completely ignored.

Nevertheless, the Greeks sensed the connection between a woman's emotionality and her genitals. The Greek word for "uterus" is *hystera,* a term we still use when we speak of the surgical excision of the uterus as a *hysterectomy.* It is because of the Greek realization of some connection between the uterus and the emotions that we still refer to an uncontrolled emotional outburst as *hysteria.* However, as we all know, men can be hysterical too, and without woman's uterus to excuse it.

THE HUMAN MALE

The male gonads consist of a pair of *testes* (tes'teez), or *testicles,* the latter term being merely the diminutive form of the former. The word is derived from a Latin term meaning "witness." There are several explanations for this, but the most logical seems to be that the testicles, when present on a man, were witnesses to his virility.

This made more sense in ancient times than it does now because it was then a fairly common practice to cut off the testicles of young slaves, a process called *castration.* This did not interfere with growth or life, but since the testes produce the hormones that in the man bring about the changes associated with puberty, a castrated boy never develops a beard, or a deep voice, and of course lacks sexual urges and is incapable of intercourse. A castrated male is a *eunuch* (yoo'nook; "guardian of the bedchamber" G), so called because he could be used as a guardian with whom a harem of women could be left in safety. An ordinary man, of normal sexual abilities, would obviously be unsuitable for the post. Fortunately the brutal and degrading practice of castration has fallen into desuetude in the civilized world.

The testes are somewhat like the ovaries in size and shape, and, like the ovaries, are originally formed in the kidney region and descend during embryonic development. In nonmammalian vertebrates they remain within the pelvis as the ovaries do. In mammals, however, they descend further. In man, about a month before birth, they leave the pelvis altogether, pushing the peritoneum ahead of them. In doing so the testes enter a pouch of skin called the *scrotum,* a Latin word of uncertain origin.

The scrotum, dangling as it does between the thighs, exposes the testicles to a temperature a few degrees lower than they would be exposed to if they were enclosed within the body. This apparently is essential for the proper development of sperm cells. Occasionally, a child is born with testicles that have not completed their descent and remain within the pelvis. This condition is *cryptorchidism* (krip-tawr′ki-diz-um; "hidden testicles" G)* and a cryptorchidist is invariably sterile, apparently through the inability of the sperm to develop at body temperature.

The interior of the testicle is one large mass of very narrow tubules that follow twisted and meandering paths. These are the *convoluted seminiferous tubules* (sem′i-nif′-uh-rus; "seed-bearing" L). The inner lining of these tubules is composed of numerous cells that are continually dividing and redividing, forming the tiny sperm cells. The head of a human sperm cell, which contains the nuclear material, is only 0.004 millimeters long. This represents a tiny cell indeed, since it is only a third of the volume of an erythrocyte, itself an unusually small cell. Attached to the head of the sperm cell is a tail that may be up to 0.15 millimeters long, but the full length of the sperm, tail and all, is considerably less than the bloated diameter of the human ovum. The sperm cell in volume is less than 1/100,000 the size of the ovum.

---

* The Greek word for "testicle" is *orchis,* and the "orchid" is so named because of the testicular shape of the tuber from which it grows. Undoubtedly few women who wear an orchid corsage know the semantic inappropriateness of doing so.

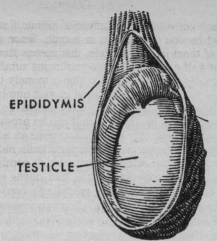

EPIDIDYMIS

TESTICLE

TESTICLE, EPIDIDYMIS,
AND CROSS SECTION
OF TESTICLE

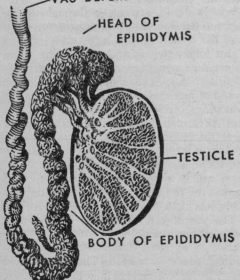

VAS DEFERENS

HEAD OF
EPIDIDYMIS

TESTICLE

BODY OF EPIDIDYMIS

TAIL OF EPIDIDYMIS

Each convoluted seminiferous tubule if straightened out would be one to two feet in length. Since there are about 800 of them in each testicle, this means that each man has about half a mile of sperm-producing tubule altogether. It is not surprising, then, that the extremely tiny sperm cells are constantly being produced in vast numbers. In contrast to the case in the female, there seems no upper limit to the number of sperm cells that can be produced, and males may continue their production to quite an advanced age.

As they are formed, the sperm cells move through the tubules, are collected in ducts which combine, and finally enter a large convoluted tube that opens at the upperpart of each testis. This is the *epididymis* (ep'i-did'mis; "upon the twins" G, the "twins" being a coy reference to the testicles, of course). The epididymis runs down the length of the testicle, slowly narrowing and straightening out, then turns upward again, as a relatively narrow and straight tube, the *vas deferens* (vas def'eh-rens; "duct carrying down" L). The vas deferens is about two feet long and carries the sperm cells up into the pelvis, around the bladder, and down toward the urethra.

At the base of the bladder are a pair of *seminal vesicles* (ves'i-kulz; "small bladder" L) which produce a fluid that is carried through a small duct into the vas deferens. From this point of junction the final section of the duct, now called the *ejaculatory duct* (ee-jak'yoo-luh-toh'ree), leads both sperm and fluid into the urethra. The genital opening thus combines with the urethral, so that the male has two openings out of the pelvis, rather than three as in the female.

In the neighborhood of the region where the ejaculatory duct enters the urethra and almost encircling the upper end of the urethra, is a chestnut-shaped organ, the *prostate gland* (pros'tayt; "to stand before" G, the reason for the name being uncertain). The prostate gland secretes a thickish fluid into the urethra which, apparently, serves to improve the motility of the sperm and to provide a more suitable environment for it. The final fluid in which the sperm is immersed is the *semen* (see'men). (The prostate can be the source of considerable trouble in later life, since there is a tendency for it to become enlarged, thus interfering with urination by its too close embrace of the urethra. It also has a distressingly large chance of developing cancer.)

The male urethra passes through the *penis* (pee′nis), which is composed of three masses of erectile tissue held together by connective tissue. The erectile tissue is a spongy mass of blood vessels for the most part. Under the proper stimulus, the arteries leading blood to the penis are enlarged and the veins leading blood away are constricted. Blood can enter the organ but not leave. The penis, engorged with blood, becomes hard and stiff.

The lowermost of the three erectile tissue masses (the one through which the urethra runs) expands at the end to form a soft, smooth, hairless, and sensitive tip to the organ. This tip is the *glans penis* (glans is the Latin word for "acorn," the name being obviously derived from the shape).

In the ordinary penis the glans is covered by a fold of skin called the *foreskin,* or *prepuce* (pree′pyoos; "before the penis" L). Among the Jews and Mohammedans, as well as among others, a religious ritual is built around the removal of the foreskin, a process called *circumcision* ("to cut around" L). When this is done, the glans penis is left visible at all times. Circumcision is becoming more common among the population generally, quite apart from religion, since it has a sanitary value. The circumcised penis is more easily kept clean. It seems quite well established that circumcision in no way interferes with sexual activity.

The penis when flaccid is about four inches long. When erect it is about six inches long. The length is remarkably constant from man to man, having little relation to the overall size of the body. The erect penis in sexual intercourse is placed within the female vagina. At the height of the activity that follows, the genital organs of both sexes are involved in a series of spasmodic muscular contractions that make up the *orgasm* (awr′gaz-um; "to boil" G). A peristalsis begins in the testes and moves along the epididymis and ducts, driving sperm cells and semen forward and shooting them out of the urethral opening and into the vagina. This process is *ejaculation* ("to throw out" L).

About three to four milliliters of semen are ejected at an

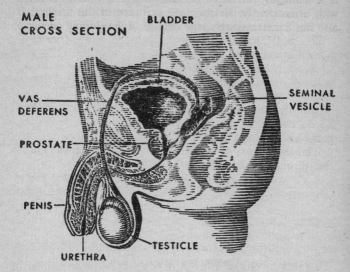

MALE CROSS SECTION

BLADDER

SEMINAL VESICLE

VAS DEFERENS

PROSTATE

PENIS

URETHRA

TESTICLE

ejaculation, and this may contain up to half a billion (!) sperm cells. The female genital organs also begin peristaltic motions, the direction of movement being inward rather than outward, so that sperm cells are drawn into the Fallopian tubes. Only one sperm cell can enter an egg cell, but the hundreds of millions of others have their uses, too. This is shown by the fact that when the number of sperm cells in a single ejaculate falls below about 150 million, fertilization does not generally take place. Though only one sperm cell may be needed for penetration, many others seem to be needed to produce enzymes in sufficient quantity to break down the protective encasement of the egg cell so that that one sperm cell might enter.

With the egg cell fertilized, the process starts which in nine months will convert it into a human body of the intricate design described in this book. From that one cell will eventually come a structure with every bone and muscle in place, every nerve and blood vessel properly laid down, every organ ready to play its part. And, eventually, the body so formed is mature enough to take part in the production of still another body.

And so on, we hope, for the indefinite future.

POSTSCRIPT

OUR LONGEVITY

TOWARD the end of 1961 America's well-known painter of primitives, Grandma Moses, died. She had begun her career of painting rather late in life, when she was almost 80. Nevertheless, she enjoyed many years in this profession, for she died at the matriarchal age of 101. I mention this because in this book on the intricacies of the human body I have laid some emphasis on the numerous ailments and disorders that can afflict it. Perhaps I ought to emphasize the reverse for a moment.

The automobile, despite its being one of mankind's most polished machines, is ancient if it lasts ten years. The human body, far more fragile, far less amenable to repair (a car's engine can be replaced; a human heart cannot), incapable of being shut down for an overhaul, and subject to far greater and more continuous difficulties, can last a hundred.

Nor need we compare the human body to inanimate objects only. How many living things that greeted the day and responded to the changing environment at the moment of Grandma Moses' birth in 1860, were still doing so on the day of her death in 1961? The list is tiny. Some trees can live centuries, and even millennia. Some giant tortoises can live up to 200 years or so. No other creatures aside from man, however, are known to top the century mark. (To be sure, there are popular stories concerning the long life span of such creatures as swans and parrots, but none of them have actually been observed to live long enough even to approach the century mark.) When Grandma Moses died, then, the world of life of 1860 had as its representative a few trees, a very few tortoises—and a few ancient men and women.

Now consider that trees live slowly, remain rooted, and stolidly stand against the buffeting of the environment. They buy longevity at the price of passivity. The giant tortoise moves—but just barely. He too buys longevity at a price: that of cold-blooded slow motion. Man is warm-blooded, however, and is as fast-moving and as deft as any creature alive. He races through life and yet manages to outlive all organisms that, like him, race, and almost all organisms that, unlike him, crawl or are motionless.

Let us restrict ourselves to the land representatives of the order to which man belongs, Mammalia. Here we can best make comparisons, for all its members are warm-blooded and all are built about the same body plan, differing only in rather minor variations.

Here it turns out that longevity is strongly correlated with size: the larger the mammal, the longer-lived. Thus, the smallest mammal, the shrew, may live 1½ years and a rat may live 4 or 5 years. A rabbit may live up to 15 years, a dog up to 18, a pig up to 20, a horse up to 40, and an elephant up to 70. To be sure, the smaller the animal the more rapidly it lives—the faster its heartbeat and breathing rate, the quicker its motions relative to its size, the more it must eat, the higher its metabolism per unit mass. For that reason, longevity becomes a more constant thing when it is measured by heartbeat rather than by year. A shrew with a heartbeat of 1000 per minute can be matched against an elephant with a heartbeat of 20 per minute and it would seem that a day in the life of a shrew sees as many heartbeats as seven weeks in the life of the elephant. In fact, mammals in general seem to live, at best, as long as it takes their hearts to beat about one billion times.

The rule is not absolute. There are exceptions, and the

most astonishing exception is man. Man is considerably smaller than a horse and far smaller than an elephant, yet he lives (or can live) to be more than 100. Nor is this the effect of modern medicine; even in days when medicine was a collection of witch doctor's superstitions, an occasional human being attained great age. On the other hand, animals, receiving the best of domestic care and medicine, wear out much more quickly than man.

Nor is this longevity the result of a metabolism that is unusually slow for a mammal. Man's heartbeat of about 72 per minute is just what is to be expected of a mammal of his size. It is faster than that of a horse and slower than that of a dog. In 70 years, which is the average life expectancy of man in the technologically advanced areas of the world, the human heart beats 2½ billion times. As for Grandma Moses' heart, that beat over 3½ billion times before she died. Considering that trees have no hearts and that tortoises (and cold-blooded creatures generally) have only very slowly beating ones, it is safe to say that the human heart outperforms all others. Certainly it outperforms other mammalian hearts by a ratio of 2½ or even 3½ to 1.

Nor can man's closest relatives, evolutionarily speaking, match him. The chimpanzee, somewhat smaller than a man, is a dotard in the late thirties. The gorilla, considerably larger than a man, is a dotard in the late forties. In terms of heartbeat they fit much more closely into the mammalian scheme than does man.

The human body, therefore, in all modesty, and from a completely objective viewpoint, is the most marvelous structure we know of. It may not have the grace of a cat or the sleek power of a horse or the tremendous strength of an elephant. It may not have the swimming ability of the seal, or the racing ability of the cheetah, or the flying ability of the bat, but it is put together for endurance and it outlives and outproduces them all.

Why should this be? Actually, no one *really* knows, and yet one can speculate—

The human being, like other mammals and, indeed, like all other living creatures above the level of the microorganism, is made up of a number of specialized cells. However, the cells of different organisms are much more alike than are the organisms themselves. The cells of any organism are made of roughly the same constituents and engage in chemical reactions of roughly the same nature. Nor are the cells very different in size. The cells of a shrew are not markedly smaller than the cells of a man; nor are the cells of an elephant markedly larger. The difference in size of the organism is due to the difference in the number of the cells, not in their individual size.

The average human body consists of about 50,000,000,-000,000 (50 trillion) cells, whereas a large elephant may consist of as many as 6,500,000,000,000,000 (6½ quadrillion) cells. A small shrew, on the other hand, may possess as few as 7,000,000,000 (7 billion) cells.

Now although the life of an organism is fundamentally based on the cell, it is more than merely the sum of the individual lives of the constituent cells of the organism. If a human being were divided into its separate cells and if those separate cells were somehow to remain alive, they would nevertheless not reconstitute a human being if merely piled together again in a random heap. The life of a multicellular organism depends not only upon the life of its constituent cells, then, but also upon the effective organization of those cells into a smoothly working whole.

Death can easily be the result of the failure of the organization. A man who suddenly dies of a heart attack or of a stroke dies with almost all his cells alive and healthily functioning. Only relatively few cells have died, but those few, by their death, disrupted the organization. An analogy can be made here between an organism and a city. New York is made up of more than seven and a half million human beings and yet they alone do not make up the city. Let a heavy snowstorm bury the city streets, or let all electric power suddenly fail, or let the city transportation workers go out on strike—and the city is in serious trouble. No one person may be directly damaged; the city's population is intact; but its organization is put out of order and that is enough to create chaos.

Let us return to the organism. It seems fair enough to suppose that the greater the number of cells composing it the more intricate is the intercellular organization necessary to keep the organism going. Yet the larger the organism the more elaborately constructed can be those organs that are primarily concerned with such organization. The two effects seem to meet in a standoff among Mammalia. The large mammals, requiring and possessing a more intricate organization, manage to maintain the function of their greater number of cells for as long as the smaller mammals do in terms of heartbeat, and far longer in terms of absolute chronology.

It might be maintained that the shrew packs as much living in its 1½ years as does the elephant in its 70, so that piling together a million times as many cells (as the elephant does, compared to the shrew) and building up a correspondingly more intricate organization achieves nothing. This, however, is not entirely so. Increased size carries with it more strength and power, a lessened fear of predators, a heightened independence of the minor vagaries of the weather. In short, a large animal is less the sport of the universe, in many ways, than a small animal is.

Again an analogy with a city is possible. A large city requires a much more elaborate organization than a small one does. Its traffic problems are much huger; its subjection to dirt and noise is much greater; its endangerment by fires, earthquakes, or other natural disasters more intense. The tremendously complicated organization enforced by its size keeps the large city viable, yet there is room to argue that it is as comfortable (or even more comfortable) to live in a small town than in a metropolis. Still, sheer size is not entirely wasted; it is the large cities of a nation that are the center of its intellectual life, of its art and culture, and even of its wealth and comfort. There is something about a Paris or a New York, that an Abbeville or a Wichita (with all due love and respect) simply cannot offer.

But surely, again returning to the organism, about man there is no question, for he is long-lived in terms of heartbeat as well as in absolute chronology. This cannot be accounted for in the cell itself, since the human cell is not markedly different from that which is to be found in other organisms. What is left, then, but to consider our intercellular organization? It seems to me that our long life is based on the fact that our intercellular organization is far more highly developed than is to be expected from our size alone. It can therefore take more buffeting and stress before breaking down than can the intercellular organization of any other living creature. It must be for this reason that we take longer to age and live longer before dying.

Then, too, we take longer to mature. We are thirteen before we are mature enough to reproduce and eighteen before we reach our full size and strength. No other mammal takes nearly as long. Surely the stretched-out length of our maturation is enforced by the fact that it takes longer for the superior human intercellular organization to develop its full powers.

Nor need the intercellular organization be considered as something too abstract to be reduced to material terms. That portion of the body most nearly concerned with organization is the nervous system (a portion of the body I did not take up in this book). The key organ of the nervous system is the brain, and if there is one human organ that is particularly unusual it is the brain. The human brain is nothing short of monstrous in size. No other land creature the size of man approaches him in brain size. The elephant has a somewhat larger brain, but that brain must exert a control over a *much* larger body than man's must.

We can conclude, then, that there are two aspects of the human body that are far, far out of line of the general mammalian pattern. One is his giant brain and the other is his long life. It would be odd indeed if there were no connection between the two.

This book has concerned itself with the parts of the human body, the separate organs composing it. It would seem that what is left—the nervous system and other organs controlling intercellular organization—makes up the better half and, in fact, makes up that which is most peculiarly and particularly *human*. It is to the intercellular organization that I shall, therefore, turn in this book's companion piece, *The Human Brain*.

INDEX

310